How the South Could
Have Won the Civil War

Other Books by Bevin Alexander

How the South Could Have Won the Civil War

The Fatal Errors That Led to Confederate Defeat

Bevin Alexander

THREE RIVERS PRESS

NEW YORK

Title page (left to right): Jefferson Davis (National Archives and Records
Administration/photograph by Mathew B. Brady), Robert E. Lee (National Archives
and Records Administration/photograph by Mathew B. Brady), Stonewall Jackson
(National Archives and Records Administration/photograph by George W. Minnes)

Published in the United States by Three Rivers Press, an imprint of the
Crown Publishing Group, a division of Random House, Inc., New York.
www.crownpublishing.com

Three Rivers Press and the Tugboat design are registered trademarks of
Random House, Inc.

Originally published in hardcover in the United States by Crown Publishers,
New York, in 2007.

Library of Congress Cataloging-in-Publication Data

Alexander, Bevin.
How the South could have won the Civil War: the fatal errors that led to Confederate
defeat / Bevin Alexander.— 1st ed.
    p. cm.
Includes bibliographical references and index.
1. United States—History—Civil War, 1861–1865—Campaigns.   2. Generals—
Confederate States of America—History.   3. Confederate States of America.
Army—Drill and tactics—History.   4. Strategy—History—19th century.
5. Command of troops—History—19th century.   6. Military art and
science—Confederate States of America—History.   I. Title.
E470.A36 2007
973.7'13—dc22   2007010816

ISBN 978-0-307-34600-1

Design by Lauren Dong
Maps by Jeffrey L. Ward

First Paperback Edition

146086900

In loving memory of my brother,

John McAuley Alexander Jr. (1919–2007),

a man of honor and a combat officer in World War II

Contents

Maps

How the South Could
Have Won the Civil War

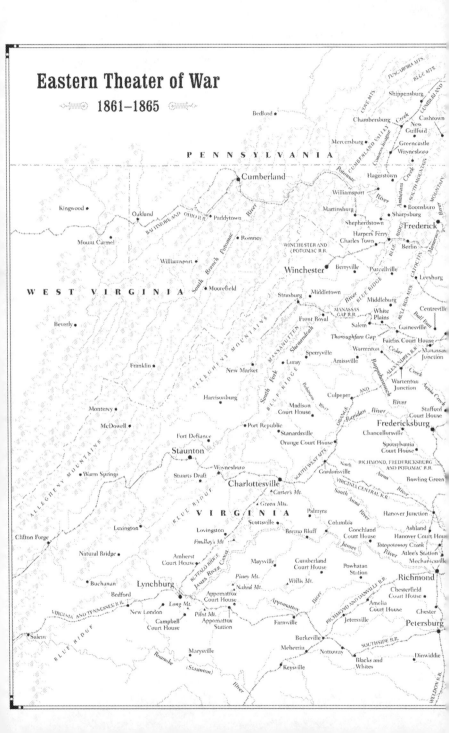

Canedaguinet Creek
Harrisburg
Carlisle
VALLEY R.R.
Trenton

Columbia
Lancaster
Wrightsville
PHILADELPHIA AND COLUMBIA R.R.
York
Susquehanna
Philadelphia
Delaware River
Camden

Gettysburg
Hanover
P E N N S Y L V A N I A
Wilmington
NEW JERSEY

Emmitsburg
Manchester
River
BALTIMORE R.R.
Salem
Taneytown
CENTRAL
AND
Westminster
Havre de Grace
Vineland

M A R Y L A N D
PHILADELPHIA WILMINGTON

BALTIMORE ROAD
AND OHIO R.R.
Patapsco River
Baltimore
Dover
Delaware Bay

Rockville
Annapolis Junction
River
Chester
BALTIMORE AND OHIO R.R.
ELK RIDGE R.R.
Severn River
Annapolis
Cape May

ALEXANDRIA
LOUDOUN &
HAMPSHIRE R.R.
Washington
D E L A W A R E
Alexandria
Upper Marlborough
River
Georgetown
Oxford
Seaford

Occoquan River
Prince Fredericktown
Chaptank
Cambridge

Port Tobacco
Patuxent River
Aquia Creek
Salisbury

Skinker's Neck
Potomac River
Rappahannock River
Port Royal
Point Lookout
Kingston

Tappahannock
Warsaw
A T L A N T I C

Mattaponi River
Henthsville
O C E A N

Lancaster Court House
Chesapeake Bay
King William Court House
King and Queen Court House
Pamunkey River
Urbanna
Cold Harbor
RICHMOND AND YORK RIVER R.R.
Fair Oaks Station
White House
West Point
New Kent
Dispatch Station
Gloucester Courthouse
Bottoms Bridge
Savage Station
Mathews Court House
York River
Seven Pines
Charles City Court House
City Point
Williamsburg
Gloucester
Prince George Court House
Yorktown
Disputanta
Surry
James
Hampton
NORFOLK AND PETERSBURG R.R.
Newport News
Ft. Monroe
River
Norfolk

0 Miles   20   40
0 Kilometers   40

Jeffrey L. Ward

Introduction

No Victory Is Inevitable

THE CIVIL WAR continues to fascinate Americans because it is uniquely *our* war. In it we were fighting ourselves. The war evoked the most intense passions Americans have ever felt, experienced all the more deeply because they were played out within our own society, sometimes within our own families. One of the questions that has been debated ever since 1865 is whether any other outcome was possible. That is, could the South have won the war?

To many observers the very question seems absurd. Given that the Confederacy had a third of the population and an eleventh of the industry of the North, the South's defeat was, according to this view, unavoidable.

But that view is wrong. This book contends that the South most definitely could have won the war, and shows in a number of cases how a Confederate victory could have come about.

Beyond the actual opportunities presented to the Confederacy, we should remember a broader fact—there is nothing inevitable about military victory, even for a state with apparently overwhelming strength. The Greeks beat the Persians at Marathon, Alexander destroyed the Persian Empire, the Americans defeated the British in the Revolution, Napoléon Bonaparte hobbled huge alliances in his early wars. In all of these cases the victor was puny and weak by comparison with his opponent.

It's true that the more powerful state usually wins. But this is because bigger states normally wear down weaker states by attrition. On average, military leadership is about equal on both sides, and this factor as a rule is not decisive. The tables can be turned, however, when a weak state produces inspired leaders. Even when great generals are only partially heeded, astonishing results can occur. For example, in World War II Adolf Hitler refused to give the German general Erwin Rommel more than scanty forces in North Africa, but Rommel nevertheless nearly brought about an Axis victory because he was head and shoulders above the British generals who came against him.

The Civil War actually was a near thing, and it was a near thing because Confederate military leadership was generally far superior to Union leadership. This superiority produced a number of Southern successes. It failed to bring about victory in the end only because the top Confederate political and military leaders failed to understand, and thus did not exploit, the opportunities offered them.

Three men more than any others determined the outcome of the American Civil War—the Confederacy's president, Jefferson Davis, and two generals, Robert E. Lee and Thomas "Stonewall" Jackson. Jackson figured out almost from the outset how to win the war, but neither Davis nor Lee was willing to follow his recommendations. It was the fundamentally different views of warfare of these three men that settled the fate of the South—not the seemingly overwhelming power of the North, and not the actions of Union commanders and their armies.

Davis was opposed to offensive action against the North. He wanted to remain on the defensive in the belief that the major European powers would intervene on the Confederacy's side to guarantee cotton for their mills, or that the North would tire of the war and give up. Even after it became plain that no European nation would come to the South's aid, Davis adhered to his conviction that the Northern people would grow weary of the war.

Lee, on the other hand, was focused on conducting an offensive war against the armies of the North. He did not see the war as a collision between the Northern people and the Southern people. He saw it as a struggle between the governments and the official armies of the two regions. Thus he wanted to confront the Union armies directly, not to strike at Northern industries, farms, and railroads, except as they served Federal armies. Without a doubt Lee was an extraordinary leader, inspiring remarkable devotion among his men and embodying the traits of honor, courage, and dedication to a cause. As a field commander, too, he was vastly superior to all of the Union commanders who came against him.[1] His mastery of battle tactics, in fact, was what permitted the South to endure four years of brutal war. But Lee's overall strategy—his insistence on frontal assaults—led to inevitable defeat. No matter how skilled a battle leader Lee was, he could never win the war by pitting the far weaker resources of the South against the tremendous economic and military power of the North. This was particularly true because revolutionary advances in weaponry had made direct assaults far more difficult to pull off and far more dangerous. Casualties in the Civil War were staggering.

Recognizing the need to adapt to the new kind of war in which they were immersed, Jackson developed a polar opposite approach. He proposed moving against the Northern people's industries and other means of livelihood. He wanted to avoid Northern strength, its field armies, and strike at Northern weakness, its undefended factories, farms, and railroads. His strategy, in short, was to bypass the Union armies and to win indirectly by assaulting the Northern people's will to pursue the war. He proposed making "unrelenting war" amid the homes of the Northern people in the conviction that this would force them "to understand what it will cost them to hold the South in the Union at the bayonet's point."[2] Significantly, William Tecumseh Sherman won the war for the North by employing precisely the strategy that Stonewall Jackson had tried but failed to get the South to follow: he conducted "unrelenting war" on the people and the property of Georgia in his march from Chattanooga to Atlanta, and from Atlanta to the sea, in 1864. This campaign broke the back of Southern resistance.

Failure to recognize the realities facing the South and disagreement over strategy are what doomed the Confederacy. That it took four bitter years of the hardest war and the most casualties in American history is due primarily to the brilliant battle leadership of Lee, Jackson, and a host of dedicated Southern officers. But wars are not won by heavy losses heroically sustained. Wars are won by ingenious plans correctly implemented. Jackson, among others, offered the South plans that would have succeeded. Davis and Lee—except at Chancellorsville—refused to carry them out.

❋   ❋   ❋

THREE DECADES BEFORE the Civil War, the great Prussian strategist Karl von Clausewitz (1780–1831) argued that in a country involved in an insurrection or torn by internal dissension, the capital, the chief leader, and public opinion constitute the *Schwerpunkt,* or center of gravity, where collapse has the greatest chance of occurring.[3] Following this theory, the Confederacy's most glittering opportunity lay not in defeating the Northern field army in Virginia but in isolating or capturing Washington, evicting Lincoln and his government, and damaging Northern industry and railroads in order to turn public opinion against the war.

British Colonel G. F. R. Henderson, the famed biographer of Jackson, made this point graphically in 1898: "A nation endures with comparative equanimity defeat beyond its own borders. Pride and prestige may suffer, but a high-spirited people will seldom be brought to the point of making terms unless its army is annihilated in the heart of its own country, unless the capital is occupied and the hideous sufferings of war are brought directly home to the mass of the population. A single victory on Northern soil, within easy reach of Washington, was far more likely to bring about the independence of the South than even a succession of victories in Virginia."[4]

Indeed, nothing would have damaged the Union war effort more than seizing eastern cities. Key targets included the rail hub of Baltimore and the metropolis of Philadelphia, which was, after New York, the largest city in the country, with some 600,000 people. The vast bulk of the North's industry was concentrated from northeastern Maryland to southern New Hampshire. Severing rail service to these areas could have prevented Union forces from invading Virginia and capturing Richmond, the Confederate capital.

Abraham Lincoln certainly recognized that the loss of Washington would have led straight to Northern defeat. That is why he insisted on employing the vast preponderance of Union power in protecting the national capital.

Likewise, Stonewall Jackson in the early days of the war realized that the Confederacy's greatest opportunity lay in striking at the North's vitals. Other Southern generals did so as well, including Joseph E. Johnston, the Confederacy's top military leader at war's beginning; Pierre Beauregard; and James Longstreet. Johnston, Beauregard, and Jackson ran up against a solid wall of resistance from President Jefferson Davis in the fall of 1861, after the victory at First Manassas, when they proposed an invasion of the North. In the spring of 1862, Jackson, Johnston, and Longstreet encountered the same refusal to act from Davis and from Robert E. Lee, who had become Davis's military adviser.

Lee, who was named commander of the Army of Northern Virginia on June 1, 1862, after Johnston was wounded, sought from first to last to fight an offensive war—that is, a war of battles and marches against the armies of the North. After Davis's rejection of invasion, Jackson turned to a new approach to warfare. Lee resisted this approach, which called for luring the Union army to attack against a strong Confederate defensive position, repelling that attack and thereby weakening enemy strength, morale, and resolve, and then going on the offensive by swinging around the flank or rear to destroy the Union army. Lee expressed his fundamental attitude about battle most cogently to his corps commander Longstreet on the first day of the battle of Gettysburg, on July 1, 1863. When Longstreet implored Lee not to assault the Union army forming up in great strength on Cemetery Ridge directly in front of him, Lee replied, "No, the enemy is there, and I am going to attack him there."

Lee betrayed his commitment to offensive warfare when he invaded Maryland in the summer of 1862. Though he justified the decision to the defensive-minded President Davis by saying he wanted simply to establish an army on Northern soil and then offer Lincoln peace on terms of Southern independence, Lee actually hoped to force a decision in the war by *attacking* Union General George McClellan's army in the North.

Stonewall Jackson urged Lee to move the Confederate army north of Washington, where it would threaten Baltimore, Philadelphia, and the capital's food supply and communications. If the Confederate army held such a dangerous position, Jackson said, the enemy would have no other option except to assault it. Lee rejected Jackson's advice once again, deciding to move west into the Cumberland Valley, far away from the center of Northern power. There he expected to fall on the Union army, not wait for it to fall on his army.

In the event, Lee did not get the chance to attack McClellan. The Union commander acquired by happenstance a copy of Lee's troop dispositions, permitting him to defeat a portion of Lee's army on South Mountain, west of Frederick, Maryland, and force Lee to abandon his broader plans. But, unwilling to give up without a fight, Lee invited a defensive battle under extremely unfavorable circumstances at Antietam on September 17, 1862. Lee stopped McClellan's attacks, but his position—backed up in a corner against the Potomac River—gave him no room to swing around McClellan's flank and defeat him. The result was a drawn battle, which presented Abraham Lincoln with the relative success he needed to issue the Emancipation Proclamation. This ensured that no European power would come to the aid of the South, thus scotching Jefferson Davis's main ambition.

Only when facing a potentially disastrous situation at Chancellorsville in May 1863 did Lee at last agree to Jackson's plan of battle. Jackson's strike down the Union western flank caused Union General Joseph Hooker's right corps to disintegrate and threw the rest of the Northern army into chaos. Jackson was just in the process of moving his corps to block the Federals' only avenue of retreat when his own troops accidentally struck him down and mortally wounded him.

Although Jackson's death handed the South a devastating blow, the Confederacy could still have won if Lee had accepted Jackson's defend-then-attack plan when he invaded Maryland and Pennsylvania a month later. James Longstreet believed he had extracted a promise from Lee to do just that. But at the very first challenge Lee faced in Pennsylvania, he reverted to direct confrontation. This led to head-on attacks on all three days of Gettysburg, July 1–3, 1863, ending with General George Pickett's disastrous charge on the third day, which wiped out the last offensive power of the Confederacy. From this point on, the Army of Northern Virginia could only strike out like a wounded lion at the forces gathering to destroy it. With Jackson dead and the South's leadership in the hands of Davis and Lee, defeat became inevitable.

But it did not have to be. This is the story of how the South could have won the Civil War. It is based not on fanciful, theoretical conjectures of what might have been but on positive recommendations proposed time after time to the South's top leaders. The concepts, recommendations, and means were at hand—at least as late as the first day of Gettysburg— for the South to emerge victorious. It did not happen because the South's primary leaders could not see the way to victory.

❊   ❊   ❊

I SINCERELY HOPE no reader will conclude that this book's title implies in any way that I am advocating some reappraisal of the Lost Cause or some nostalgic longing for what is gone with the wind. To use an old Southern phrase, we are well shut of our selfish aristocrats, who tried until the last to hold on to the unpaid labor of slaves so as to get and keep their wealth. The elimination of slavery and the aristocrats who fed on it was a glorious and long-overdue advance.

This book is about something entirely different. I hope to move a step beyond the comment of the great New York author Damon Runyon (1884–1946), who wrote, "The race is not always to the swift, nor the battle to the strong—but that's the way to bet." My purpose is to show that, despite the odds, wars are won by human beings. When superior military leaders come along and political leaders pay attention to them, they can overcome great power and great strength. That is a lesson we need to remember today.

1

"There Stands Jackson
Like a Stone Wall"

O NLY THREE MONTHS after Confederate troops fired on Fort Sumter in Charleston harbor on April 12, 1861, Confederate leaders had to confront an invading Union army under General Irvin McDowell. This army was seeking a decision in the war by challenging a Southern army under Pierre Beauregard of Louisiana, the officer who had fired on Sumter. Beauregard's troops were lined up along the banks of Bull Run, a bold stream some twenty-five miles south of Washington in northern Virginia, near the little town of Manassas.[1]

There had scarcely been time for the South to organize military units, let alone come up with a strategic plan for achieving independence. The North did have a plan, however, and McDowell's march on Bull Run was a part of it. The other parts, devised by Winfield Scott, the septuagenarian commander of the U.S. Army, called for throwing a sea blockade around the South and sending an army down the Mississippi River.[2]

McDowell's movement was to be the first stage of a drive to Richmond, the Confederate capital. But lots of people in Washington thought the Southern army would be swept away by the larger and better-equipped Union forces at Bull Run, and that would be the end of it. The chastened Southern states would come back meekly into the Union and the country could go on about its business.

There was solid reasoning behind this view. The South had few weapons and fewer machine shops to build them, and the army Beauregard commanded had been built from scratch and seemed likely to disintegrate at the first tap. The North, however, had been able to pour unlimited money and materials into the new army that Abraham Lincoln had called to put down the "rebellion."

But things in the South were not quite as seen from Washington. Lincoln's decision to invade unified the Southern people and gave them tremendous resolve to resist. Volunteers flocked to the colors all over the South, and the Confederacy's leaders, generally with more training in military affairs than the Northern chiefs, set out to create an army. They searched for every qualified man and tried to give him an assignment commensurate with his skills and promise, and the South's needs.

Typical of the rapid response was the action of Virginia. Shortly after the state seceded on April 17, 1861, Governor John Letcher telegraphed the superintendent of the Virginia Military Institute at Lexington and ordered the corps of cadets to report to Richmond immediately to teach the flood of recruits gathering there how to march and wheel in formation—a skill the cadets knew to perfection. The superintendent chose Major Thomas J. Jackson to command the movement of the 176 young men. A West Point graduate, thirty-seven years old, thin, moderately tall, taciturn, deeply committed to his Presbyterian faith, and a hero of the Mexican War of 1846–48, Jackson had left the army in 1851 to become a professor of artillery tactics and mechanics at VMI. By April 26, Governor Letcher had commissioned Jackson as a colonel and sent him on from Richmond to command at Harpers Ferry on Virginia's frontier, at the northern exit of the Shenandoah Valley.

On April 18, shortly before Jackson and the VMI cadets set off for Richmond, Robert E. Lee, a highly regarded colonel in the U.S. Army, was in Washington, having been summoned there by his old boss, General Scott, on whose staff Lee had served with distinction during the 1847 march on Mexico City in the Mexican War. Lee, scion of a famous Virginia family, was fifty-four years old, fairly tall, extremely handsome, unfailingly courteous to everyone high and low, and an honors graduate of West Point, where he had been superintendent from 1852 to 1855. Francis P. Blair Sr., a publisher and close associate of President Lincoln, met with Lee and informed him that Lincoln had offered him command of the U.S. Army. Lee declined, saying he could take no part in the invasion of the South. Though deeply distressed at the division of the country, Lee felt he had to offer his sword to his native state. On April 20, he resigned his commission, left his estate at Arlington, just across the Potomac from the capital (present-day site of Arlington National Cemetery), took the train to Richmond, and accepted Governor Letcher's appointment as major general and commander of Virginia forces.

His job did not last long. On May 2 Virginia ratified the Confederate Constitution and joined the new nation. When President Jefferson Davis moved the Confederate capital from Montgomery, Alabama, to Richmond on May 29, Lee became an unofficial military adviser to the Confederate president (who, like the Union president, was the constitutional commander in chief). Lee recommended to Davis that Beauregard be assigned to command the troops at Manassas, but he had no role in the upcoming battle.

Meanwhile, the Confederate War Department assigned one of its highest-ranking officers, Brigadier General Joseph E. Johnston, a Virginia native, to command at Harpers Ferry, superseding Thomas Jackson, who was assigned to lead the 2,600 men of the new 1st Virginia Brigade. Johnston felt Harpers Ferry was indefensible, and moved his little army twenty-five miles southwest to Winchester.

❋   ❋   ❋

THE MILITARY SITUATION in July 1861 was as follows: Irvin McDowell had a Federal army of 37,000 men at Alexandria and vicinity facing P. G. T. Beauregard's 22,000 men at Manassas, while another Union army of 15,000 men under Robert Patterson had moved south of Harpers Ferry to challenge Joseph Johnston's 11,000 men at Winchester. Washington leaders feared that Johnston might burst out of the Shenandoah Valley and attack the capital while McDowell was dealing with Beauregard at Bull Run.

On July 16, McDowell started on a thirty-mile march from Alexandria west to Bull Run. He took 32,000 men, leaving 5,000 in defensive works guarding Washington. Word of the march leaked out the next day. Patterson's clear duty was to keep Johnston bottled up in the Shenandoah Valley. Winfield Scott had assured McDowell that Patterson would do just

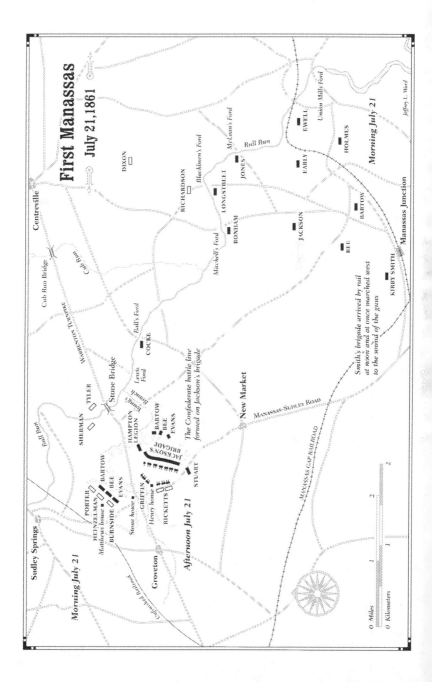

First Manassas
July 21, 1861

Sudley Springs

Morning July 21

Centreville

Cub Run Bridge

Cub Run

WARRENTON TURNPIKE

Ball's Run

Ball's Ford

TYLER

SHERMAN

Stone Bridge

Lewis Ford

COCKE

Young's Branch

HAMPTON LEGION

BARTOW
BEE
EVANS

The Confederate battle line
formed on Jackson's brigade

JACKSON'S BRIGADE

STUART

New Market

MANASSAS-SUDLEY ROAD

PORTER

HEINTZELMAN

BARTOW

BEE

Matthews house

BURNSIDE

EVANS

GRIFFIN

Stone house

Henry house

RICKETTS

Groveton

Afternoon July 21

Orange & Alexandria Railroad

Morning July 21

DIXON

RICHARDSON

Blackburn's Ford

Bull Run

McLean's Ford

Mitchell's Ford

BONHAM

LONGSTREET

JONES

JACKSON

BEE

EWELL

Union Mills Ford

EARLY

HOLMES

BARTOW

KIRBY SMITH

Morning July 21

Manassas Junction

Smith's brigade arrived by rail
at noon and at once marched west
to the sound of the guns

MANASSAS GAP RAILROAD

0 Miles   1   2
0 Kilometers   1   2   3

Jeffrey L. Ward

that. Unfortunately Patterson's duty was not clear to Patterson himself, who was sixty-nine years old and had no experience in independent command. Johnston made some threatening moves against Patterson's Union army at Bunker Hill, ten miles north of Winchester. Thoroughly intimidated and believing Johnston might cut him off from his supplies, Patterson hurried his army northeastward to Charles Town on the very day that McDowell commenced his march to Bull Run. This put him twenty miles from Winchester and released Johnston from the vise.

The first principle of war is to concentrate superior force at the decisive point. In July 1861 this point was along the banks of Bull Run. McDowell and Patterson together commanded more than 50,000 men. The South, even if Johnston and Beauregard could get together, had only 33,000. In the days of single-shot infantry weapons, numbers counted for very much, because adding numbers was the only way firepower could be increased.

General Scott knew his forces should be concentrated. He also knew it would be impossible for Johnston's Confederate forces to march twenty-five miles to Harpers Ferry, cross the Potomac River, then drive the fifty-five miles to Washington before a decision could be reached at Bull Run. He explained to President Lincoln that if Johnston attempted this, he would be operating on "exterior lines," or on a roundabout approach, whereas Patterson, in marching to reinforce McDowell, would be moving directly, thus operating on "interior lines." But this was too arcane a concept for Lincoln. He reflected rather the panic he felt all around him, and insisted that Patterson be kept in the Shenandoah Valley to hold Johnston at bay.

Events were to show that Beauregard and Johnston were in the second tier of Confederate commanders in terms of ability. But they saw that Lincoln's anxiety had dispersed Union forces, giving them the opportunity to make up in activity for their lack of strength. They rushed to combine their forces against only a part of the enemy and defeat this part before having to deal with the other part.

At 1:00 A.M. on Thursday, July 18, 1861, a wire from the War Department in Richmond alerted Johnston that McDowell was on the march.

Move quickly, the wire ordered, to Manassas if practicable. Johnston set his army in motion the fifty-seven miles to Manassas.

Johnston sent his flamboyant cavalry chief, J. E. B. ("Jeb") Stuart, off at a gallop with his 1st Virginia Cavalry Regiment toward Charles Town to create confusion and seal off Patterson's avenues of reconnaissance. For that entire day and for most of the next, Patterson was unaware that Johnston's army had disappeared. When he did learn, it was too late. One adept, fast move by Johnston had wiped Patterson's whole army from the order of battle. In the collision to come, Patterson's forces might as well have been on another planet.

Thomas Jackson's 1st Virginia Brigade led the march from Winchester across Ashby's Gap in the Blue Ridge Mountains (through which U.S. Routes 17 and 50 now run). At the little village of Paris, on the eastern slopes of the Blue Ridge, twenty miles from Winchester, Jackson's exhausted troops collapsed for the night. An officer reminded Jackson that not enough pickets had been posted. Jackson replied, "Let the poor fellows sleep. I will guard the camp myself." And Jackson, through the summer night, stood sentry over his sleeping men.

Next morning, the brigade marched six miles down to Piedmont Station (now Delaplane) on the Manassas Gap Railroad and entrained for Manassas Junction, thirty-four miles away. This was just the second time in history that troops had been moved by rail to battle. Only in 1859 in a brief conflict between France and Austria in Italy had it been done. Armies on both sides would increasingly rely on railroads not only to move troops quickly and in large numbers—something never possible before—but also to supply these armies at distant points. The old order of relying on magazines, or supply depots, maintained only a short distance from troops, passed away. The umbilical cord of armies now became the rail line. If it was intact, armies could survive wherever they were located. If it was severed, armies had to repair the cut, move, or die.

But none of this was in the minds of the young soldiers of Jackson's brigade as they rode across the rolling Virginia countryside on July 19, 1861. The trip took eight hours, and it was a lark. At every stop enthusiastic crowds swarmed around the cars, girls in their finery waved and flirted, housewives showered them with food, and everybody sang patriotic songs such as "Dixie" and "The Bonnie Blue Flag."

Jackson announced his brigade's arrival to Beauregard at his headquarters at the Wilmer McLean farmhouse, on a shady knoll half a mile below Bull Run.[3] He told Beauregard that Johnston's remaining three brigades were coming by rail behind him.[4]

But these troops were in the process of becoming the victims of military history's first railway traffic jam. The three brigades reached Piedmont Station during the afternoon of July 19, but the railroad had no extra employees to bring in. Train crews were able to get only Francis S. Bartow's brigade of 1,400 men to Manassas by dawn on July 20, while Johnston himself arrived with 1,000 men from Barnard E. Bee's brigade during the day on July 20. On the day of the battle, Sunday, June 21, 2,000 men from E. Kirby Smith's brigade reached the Manassas station at noon and marched directly into the fight. The last 2,500 infantry of Johnston's army did not arrive until after the battle was over.

Beauregard conceived plans that were far too airy to be carried out. On July 19, he asked Johnston to throw his army on McDowell's right flank at Centreville, on the Warrenton Turnpike (now U.S. Route 29) some three miles above Bull Run, while Beauregard attacked northward from the stream. Combining two parts of an army on the field of battle is an extremely difficult task even for experienced forces, which the Confederate troops were not. To be successful, this approach requires flawless execution, no unanticipated barriers, and a cooperative enemy. Johnston sensibly rejected Beauregard's idea and insisted on combining the two forces behind Bull Run.

Beauregard had organized his army poorly and had placed it in a dangerously exposed position. While McDowell had formed his army into five divisions, a structure whose effectiveness had been proved in the Napoleonic wars more than half a century before, Beauregard had rejected divisions and left his smaller force divided into six independent brigades, with a number of separate regiments and artillery batteries. This made the command structure far too complex for him and his untrained staff to handle efficiently.

Beauregard had set up a line of battle stretching nearly seven miles along the somewhat crooked course of Bull Run, much too extensive for the troops he possessed. He had concentrated the bulk of his forces—five brigades, or about 16,000 troops—in the three miles between Union Mills Ford on the right, or east, and Mitchell's Ford to the west. Two other fords, Blackburn's and McLean's, were located along this stretch of the stream. Beauregard had another 5,000 men: Philip St. George Cocke's brigade and Nathan G. (Shanks) Evans's small detachment (the 4th South Carolina Regiment and the 1st Louisiana "Tigers" battalion). But he positioned these troops too far to the west to be available for any coordinated action; Cocke stood at Ball's Ford, a couple of miles west of Mitchell's Ford, while Shanks guarded the Stone Bridge over Bull Run on the Warrenton Pike.

In the terminology of the time, both of Beauregard's flanks were floating "in the air," meaning they rested on no solid defensive position such as a stream or a mountain. They were set up to defend Bull Run to the north, not to hold against an enemy movement from either the west or the east.

When Jackson's brigade arrived on July 19, Beauregard placed it in reserve behind Mitchell's Ford.[5] On July 20, when the brigades of Bartow and Bee arrived (along with Johnston himself), Beauregard stationed them in reserve behind Blackburn's Ford.

Thus on the eve of battle, the only reserve forces with any hope of quick movement to the left or west—the most obvious point of danger—were Jackson's, Bee's, and Bartow's brigades. And they were still four to five miles away from Evans at the Stone Bridge.

Beauregard's plan of battle was based on his conviction that McDowell would attack at Blackburn's Ford. He expected James Longstreet's brigade, positioned at this ford, to stop the attack, and two brigades to the west—Milledge L. Bonham's at Mitchell's Ford and Cocke's at Ball's Ford—to hold. At the same time, he expected three other brigades—Richard S. Ewell's, D. R. Jones's, and Jubal A. Early's—to cross Bull Run, execute a wide movement around the Union army, and seize Centreville, three miles north of the stream, and Fairfax Court House, eight miles east of Centreville. As soon as this turning movement was under way, Bonham and Longstreet were to join in.

The flaws in this plan were not only that it required McDowell's full co-operation but also that Beauregard had not given the five brigades on the east clear instructions as to what they were to do and had not appointed an overall commander for this wing. Moreover, he had left his entire left flank protected only by Shanks Evans's small force at the Stone Bridge.

Unfortunately for Beauregard's planning, McDowell did not intend to attack at Blackburn's Ford but expected to go around the Confederate right, or eastern, flank, and roll up Beauregard's army from east to west.

When the leading element of McDowell's army, a division under Daniel Tyler, arrived on the north side of Bull Run facing Blackburn's and Mitchell's fords around noon on July 18, McDowell told Tyler to make threatening gestures at the fords ahead but not to bring on a general engagement. Meanwhile, McDowell rode off east of Union Mills Ford to search out a suitable place to cross Bull Run and sweep around the Confederate army.

After he departed, Tyler ordered Colonel Israel B. Richardson's brigade to scour the woods along the stream. This brought on a chaotic engagement with James Longstreet's brigade, which had been defending Blackburn's Ford. The engagement had little purpose and accomplished virtually nothing. Both sides lost only a few men. Afterward an artillery duel filled the valley with gunsmoke. This threw off the aim of the gunners, and the shelling did little damage.

McDowell found he could not turn the Confederate eastern flank. The country was too wooded and broken, with too few roads. Also, based on Richardson's confused collision with Longstreet, he concluded (incorrectly) that the Confederates were too strong along the stream to be assaulted head-on. McDowell ruled out another alternative, to strike directly down the Warrenton Pike from Centreville, seize the Stone Bridge, and then turn eastward against the Confederate army along Bull Run. Reconnaissance stated that the Confederates had mined the Stone Bridge and would blow it the moment Union troops tried to cross, and that the bridge was protected by large cannons and heavy abatis (logs piled up as a barrier). None of this was true. Evans had placed only a few logs on the roadway, had not mined the bridge, and had only a couple of light six-pounder guns, which could have done little to stop an attack.

McDowell decided instead to shift his strike force of two divisions west of the Stone Bridge, cross Bull Run at a suitable ford, then descend on the left flank of the Confederates. His engineers found an easy ford at Sudley Springs, two miles west of the bridge as the crow flies, but six miles by a farm road branching off the Warrenton Pike. From Sudley Springs a road led straight to Manassas, crossing the Warrenton Pike a mile beyond the Stone Bridge. The strike force was to attack the Stone Bridge from the rear while three brigades of Daniel Tyler's division were to make a demonstration at the bridge (that is, to make threatening gestures but not launch a full-out attack). Once the bridge was carried, Tyler was to join the other two divisions. Together they would shatter the Confederate army by sweeping behind its left flank.

Two Union divisions set off in the early hours of July 21 to make the circuitous march to Sudley Springs. The movement was entirely hidden from the Confederates. Thus it seemed that McDowell was ready to use one of the principal means by which an army can achieve victory: surprise. The main benefit of surprise is to permit one's force to arrive at a critical place, but one unguarded or lightly guarded by the enemy. In this case the critical place was the undefended left flank of the Confederate army. The great Confederate cavalry raider Nathan Bedford Forrest explained the concept well. The secret of warfare, he said, "is to get there first with the most."

But then McDowell committed a blunder of colossal proportions. For his turning movement to be successful, he needed to ensure that the Confederate troops lined up along Bull Run could not detach forces to meet the new threat on the flank. To do that required a fierce and powerful attack that would hold the enemy in place. The holding attack had been Napoléon Bonaparte's most successful method of tactical victory. Military schools and writers in Europe and America had been studying his practices for decades, especially his victory at Castiglione in northern Italy in 1796, where Bonaparte had demonstrated step by step the necessity of combining a powerful holding attack with a descent on the enemy's flank or rear.

But McDowell ignored all of the received wisdom. Instead, he ordered Colonel Israel Richardson's brigade, and Colonel Dixon S. Miles's two-brigade division north of Mitchell's and Blackburn's fords, not to launch a fierce attack.[6] The Union general compounded his failure by telling Daniel Tyler not to attack Shanks Evans at the Stone Bridge, either. Thus, when the time came, Evans was able to move easily.

❖   ❖   ❖

CONFEDERATE COMMANDER P. G. T. Beauregard, still expecting an attack at Blackburn's Ford, was stunned to learn that Tyler's Union forces had arrived around 6:30 A.M. on July 21 at the Stone Bridge and begun pounding the ground behind it with artillery fire.

So despite Joseph Johnston's brilliant move in bringing his army to Bull Run and canceling out Robert Patterson as a factor in the battle, Beauregard still had failed to concentrate against the Union army. McDowell, on the other hand, was throwing the bulk of his army against the weakest and most exposed part of the Confederate army. Most of Beauregard's force was positioned far from the crucial left flank, and only Johnston's recently arrived brigades had any immediate hope of supporting Shanks Evans. Beauregard's planning, therefore, had provided little that could make success possible for the Confederates. Their only hope lay in the potential errors of McDowell.

Beauregard directed Bee's and Bartow's brigades to rush to Evans's assistance, while he sent Jackson's brigade to cover the largely undefended ground along Bull Run between Philip St. George Cocke's brigade at Ball's Ford and Milledge Bonham's brigade at Mitchell's Ford. Evans quickly realized that Tyler was not going to assault the Stone Bridge, but he soon learned of the real danger facing his small force.

At that time, Captain Edward Porter Alexander of Georgia was positioned on Wilcoxen Hill, about six miles east of the Stone Bridge. Twenty-six years old, an 1857 graduate of West Point, and an instructor in engineering there, Alexander had been part of a team that developed a system of sending military messages over a distance, using telescopes or field glasses to view flags or torches making dot-dash or "wig-wag" signals.[7] Beauregard had ordered him to stay at the Wilcoxen signal station, keep his eye on the whole field, and report anything he might discover.

Alexander was most reluctant to undertake the assignment, but he duti-fully set to work.

At 8:45 A.M., his glasses caught a flash of light beyond the Stone Bridge. He recognized that this was the reflection of the morning sun on a brass cannon. Alexander began an intensive search of the area and dis-covered the glitter of bayonets and musket barrels. A Union column more than half a mile long was crossing Sudley Springs Ford. Its length indicated it was at least a division in size.

Alexander realized a crisis was at hand. Even before notifying Beaure-gard, he flashed a message to Evans via another Confederate signal sta-tion, on the Van Pelt farmstead, just a few hundred yards northwest of the Stone Bridge: "Look out for your left; you are turned."[8] Alexander's mes-sage confirmed a report Evans received at the same time from one of his pickets that a Federal column was moving toward Sudley Springs Ford.

Beauregard's proposed attack on the eastern Union flank instantly vanished, but he forgot entirely to cancel it.[9] Beauregard did, however, send a courier to Thomas Jackson and a 600-man battalion under Wade Hampton of South Carolina, just arrived from Richmond, with orders to go to the aid of Evans.

A less courageous officer than Shanks Evans would have withdrawn in the face of such overwhelming odds. But Evans, informing Cocke on his right, and leaving about 200 men to guard the bridge, moved the rest of his little force, with his two small field pieces, onto Matthews Hill to his immediate west and just across the Warrenton Pike.[10] There, a little more than a mile from Sudley Springs Ford, Evans's men spread out in a de-ceptively long line to give the impression of strength. They were partially protected by a grove of trees with a clear field ahead of them over which the Federals would be obliged to march. It was 9:00 A.M.

At 9:15 A.M. the lead brigade of the Union strike force, commanded by Colonel Ambrose E. Burnside, climbed onto the slopes of Matthews Hill and emerged from thick woods into the open field. Both the 1st Rhode Island and the 2nd Rhode Island regiments tried to attack, but each met fierce resistance from Evans's entire line. Before trying to as-sault again, Colonel David Hunter, commanding the lead division of the

Federal strike force, brought up other brigades to form into a long and more powerful line. While this was happening, Hunter suffered a severe neck wound.

Colonel Andrew Porter took command. But in the confusion surrounding the wounding of Hunter, Evans sent the 1st Louisiana "Tigers" in an assault directly on the Union center. The charge had no chance of routing the much larger enemy force, but its ferocity delayed the advance for more than an hour, giving time for Confederate reinforcements to arrive.

Evans was helped by timid and uncoordinated Federal efforts. Porter sent forward one regiment at a time—just 700 to 900 men—permitting Evans's small force to concentrate against the Union attack. Tactical doctrine called for an attacking force to march directly on the enemy and shoot him down at close range. This required strong discipline and iron resolve, and was extremely difficult for the green Union troops to accomplish, especially if they were isolated and unsupported. Their practice was to "fire and fall back" to cover.[11] If Porter had formed two brigades in a double line, he could have enveloped both flanks of Evans's force and crushed it inside of ten minutes.

When Barnard E. Bee's brigade finally arrived as Shanks Evans's reinforcements, Bee ordered his men to drop to the ground to avoid the bullets hurtling through the air. He told them to stand and fire only on command, then fall again to the ground to reload. Francis S. Bartow's brigade arrived soon after and went into line to the right of Bee, the men also lying down to escape the bullets and the artillery rounds, and rising to fire on command.

Total Confederate strength was about 3,600 men, a fourth that of the two Federal divisions. The entire Confederate line rose to charge two Federal artillery batteries that had emplaced next to a grove of trees, but the Rebels' small field pieces were no match for the Union guns. Only two Confederate regiments—the 8th Georgia and the 4th Alabama—got very far in the storm of fire that greeted them, and even they soon had to withdraw.

By now the second division of the Union strike force, under the command of Samuel Peter Heintzelman, had reached the field and formed up on Porter's right. The attack threatened to envelop the Confederate left, but the fire from the Confederate line caused heavy losses and great disorganization among the Union forces.

Meanwhile, another Union brigade from Tyler's division, commanded by Colonel William Tecumseh Sherman, advanced on the Rebel rear on Matthews Hill. Sherman's brigade had marched upstream about a mile and crossed Bull Run at a ford Sherman had discovered. This unexpected approach unnerved the Confederates, who began to retreat, the withdrawal becoming increasingly disorganized and panicked. Around 11:30 A M. they moved east across the Warrenton Pike and onto Henry House Hill, an undulating, largely open field rising to a low plateau.

General McDowell rode along the Federal line descending Matthews Hill shouting, "Victory! Victory! The day is ours!" The Union army was spreading out to envelop the Confederate left flank, and it seemed only a small task for the Federals to sweep up and over Henry House Hill. If the Rebels lost this hill, there could be no rallying point to the east, and the Confederate army would collapse in chaos.

※　　　※　　　※

So far, events had gone well for McDowell. He had been able to concentrate more than half his army on the Confederate left, while Beauregard had managed to get less than a third of his forces there.

As Evans, Bee, and Bartow were streaming off Matthews Hill, Wade Hampton's South Carolina battalion came forward to the Warrenton Pike to cover the retreat. Although he lost a fifth of his men, Hampton held up the Federal advance for an hour before falling back.

There remained one solid Confederate position on the western flank. It was held by the brigade of Thomas J. Jackson. While guarding Bull Run some two and a half miles east of Henry House Hill, Jackson had heard the sound of battle. Without waiting for orders, he had moved off, sending a courier to notify Bee that he was coming.

Seeing that the Federals were driving the Confederates off Matthews Hill, Jackson stopped on the crest of the rolling, expansive plateau of Henry House Hill, which sloped down to the Warrenton Pike. Though much of the pasture was open, the middle of the crest and the fall to Bull

Run on the right were covered with a dense thicket of young pines. The left side of the crest was more open and was bounded by the sunken road from Sudley Springs to Manassas. The modest wooden one-story house of Judith Henry was just off this road halfway between the crest and the pike.[12] The crest, Jackson recognized, was a comparatively good defensive position. He halted his brigade, formed it up on the reverse slope, and waited.

Soon afterward, Captain John D. Imboden came by, leading three guns of his Staunton (Virginia) Artillery to the rear. He had been firing from a position halfway down Henry House Hill but decided he was too exposed. Imboden told Jackson of the heavy beating he had taken from the Federal guns.

"I'll support your battery," Jackson told him. "Unlimber right here."

Imboden said he had only three rounds of ammunition left and suggested that he go to the rear to fill his caissons.

"No, not now," Jackson replied. A new defensive line had to be established. The guns, ammunition or not, would give the impression of strength. Imboden turned his guns around and pointed them toward the enemy. He was soon joined by the guns of the Rockbridge (Virginia) Artillery, attached to Jackson's brigade, and two pieces of Captain Philip B. Stanard's Thomas Artillery from Richmond.

Jackson now had nine guns in a formidable line along the crest of the hill. Jackson's battle line stretched out for several hundred yards but remained hidden below the crest. His right was protected by the pines and a steep slope leading up from Bull Run, but his left was not anchored on good defensible ground. Seeing this, Jackson found J. E. B. Stuart and asked him to post his 1st Virginia Cavalry as flank guard.

Jackson directed his infantry to lie down in the rear of the artillery in a shallow depression on the reverse slope. There, largely protected from Union fire, they could still sweep the crest of the hill if Union troops reached it and silhouetted themselves against the sky.

Jackson's men lay in their position as artillery rounds roared over their heads. A few men were killed or wounded by exploding shells. The smoke, the noise, the explosions, and the casualties frightened the soldiers, but Jackson calmed them, riding slowly back and forth along the line. "Steady, men, steady!" he repeated over and over. "All's well."

Once, when he had raised his hand, a bullet or a piece of shrapnel struck the middle finger on his left hand, fracturing it. But Jackson, wrapping a handkerchief around the wound, told bystanders, "Only a scratch, a mere scratch."

The broken regiments of Evans, Bee, and Bartow had taken cover in a ravine near Young's Branch, a tributary of Bull Run, on Jackson's right front. They had lost all order and cohesion. Upon the withdrawal of Wade Hampton's battalion, Federal troops began preparing to ascend Henry House Hill, heading directly toward Jackson's position.

A solitary horseman rode up from Young's Branch toward the Confederate line. It was General Bee of South Carolina, six feet tall, black eyes, dark mustache, and long hair. He halted and asked what troops were these. A soldier pointed to Jackson. As Bee approached, Jackson recognized him at once. They had been friends at West Point.

"General," Bee reported, "they are driving us!"

Jackson responded: "Sir, we will give them the bayonet!"

Jackson's resolve filled Bee with renewed confidence. He galloped back to the ravine, where his officers were trying to reorganize the units and set up a defensive line.

Bee rode into the middle of the throng, pointed his sword toward the crest of Henry House Hill, and shouted: "Look, men, there stands Jackson like a stone wall! Rally behind the Virginians!" The effect was quick. Helped by Beauregard and Johnston, who had arrived on the field, the troops re-formed and moved up alongside Jackson's brigade. There they did good service, though Bee was killed during the battle.

Beauregard and Johnston also were working furiously to get men and cannons to Jackson's line. First to arrive were parts of Cocke's brigade from Ball's Ford. Jubal Early's brigade was on the march from its position along Bull Run, as were two regiments of Milledge Bonham's brigade. Beauregard sent out orders for Richard Ewell's brigade, on the extreme right, to hurry forward. Kirby Smith's brigade had arrived by rail at Manassas station at noon and was marching to the sound of the guns.

The Confederate position slowly stiffened and began to spread out. Reinforcements went into line alongside Jackson. His brigade—forever after to be known as the Stonewall Brigade—had become the concentration point of the Confederate army.

⁂   ⁂   ⁂

IRVIN MCDOWELL STILL had a chance to win the battle. Even when the Southern reinforcements came onto the line, Beauregard had only about 7,000 men and sixteen guns, and these of light caliber; McDowell still had about 19,000 men on this left flank, and twenty-four guns, mostly heavier rifled pieces. Many hundreds of faint-hearted Confederates were streaming to the rear, abandoning their comrades, and spreading wild tales of the imminent collapse of the Confederate army.

At this critical juncture, however, McDowell made a fatal tactical mistake: he attempted to carry the Confederate line with less than one-half of his force—four brigades. McDowell had left Oliver O. Howard's brigade in the rear as a guard, a wholly unnecessary precaution. He at last called it forward, but it arrived too late to affect the issue. He allowed Burnside to beg off any further action with the claim that he was out of ammunition. Erasmus Keyes's brigade, which had followed William Tecumseh Sherman's across Bull Run, halted in the valley below the Stone Bridge and, inexplicably, did nothing. The last brigade of Daniel Tyler's division, that of Robert C. Schenck, did not even cross Bull Run and was entirely useless.

Confederate captain Porter Alexander remarked about Keyes's brigade, "Had it advanced upon the Confederate guns, or had it communicated with Schenck's brigade on the north side of Bull Run, and the two moved on Cocke's brigade at Ball's Ford—had it, in short, tried *anything,* it might have accomplished important results."[13]

Because of this failure in command, McDowell could bring only about 9,000 men against the Confederate army gathered around Jackson's brigade. At the moment, this Southern army was still smaller than McDowell's force, but—because of Jackson's wisdom—it occupied a sound defensive position. Moreover, two strong brigades, Jubal Early's from Bull Run and Kirby Smith's from the railroad station, were coming up soon to reinforce Beauregard. When they arrived, the Confederates would be numerically superior.

McDowell now committed two more errors that ruined any possibility of victory. First, instead of moving around the Confederate army's relatively open flank along the Sudley-Manassas road, where he would have encountered fewer defenders and less fire, he tried to push up the slope of Henry House Hill, directly into the massed strength of the Confederates. Second, he repeated the same blunder that had caused his advance on Matthews Hill to be so slow—he sent individual regiments, one after another, into the attack, instead of forming up his brigades in a single, massive assault.

For the next two hours—while Confederate reinforcements were rushing for the field—one Union regiment after another formed up at the bottom of the hill in the covered lower reaches of Young's Branch, then advanced toward the Confederates on the crest. As soon as the heads of the soldiers appeared over the front edge of the plateau, they were met with a hail of fire. The Union soldiers returned a volley or two and then ran back down the hill until they found cover again. Some regiments repeated this a number of times, but none ever made a lodgment.

At about 3:00 P.M., seeing that his infantry was getting nowhere, Mc-Dowell ordered two batteries of regular army artillery, twelve rifled cannons, commanded by Captains Charles Griffin and James B. Ricketts, to move forward to the Henry house, only about 330 yards from Jackson's left flank. Supporting them were two infantry regiments, the 11th New York (whose uniforms mimicked the colorful Zouave attire worn by Algerian recruits in the French army) and the 14th New York.

McDowell's aim was to break a hole in the Confederate line with canister—a deadly cloud of metal pellets and fragments fired from cannons. As the two batteries moved boldly into position and began to fire, the two New York supporting regiments took cover behind the deep bank bordering the sunken Sudley-Manassas road near the Henry house.

Shortly after they arrived, 150 troopers of Jeb Stuart's 1st Virginia Cavalry rushed out of the woods on Jackson's left and slammed, sabers flashing, into the Zouave regiment. The Zouaves opened ranks as the horsemen dashed through, shooting down nine troopers and eighteen horses. Even so, the Confederate cavalry charge unnerved the men of the two New York regiments.

Soon after the cavalry returned to the woods, the 400-man 33rd Virginia, also on Jackson's left, without orders charged the Federal guns. Captain Griffin saw the Rebels coming and prepared to shatter them with canister. But Major William F. Barry, McDowell's artillery chief, insisted that it was the 14th New York coming to support the guns. A moment later the regiment proved its identity by a deadly volley that cut down forty gunners and seventy-five horses and badly wounded Ricketts.

This was too much for the two New York infantry regiments. They instantly broke and fled in confusion to the rear. Griffin summoned the few remaining men in the batteries and was able to draw off three of the guns. Nine guns remained abandoned, however, and the 33rd Virginia rushed onto the position with every intention of holding it.

Colonel William Tecumseh Sherman moved his brigade up the Sudley-Manassas road, using its high bank as cover, with the aim of recovering the guns for the Union army. When his lead element, the 2nd Wisconsin, got abreast of the position, Sherman ordered it to turn left and assault. The 33rd Virginia at the guns opened heavy fire as the regiment came into view, forcing it to retire back to the road. Now the 79th New York had closed up, and Sherman ordered it to attack. By this time, a Confederate cannon had found the range and was dropping heavy and accurate fire on the advancing regiment, while the Virginians delivered strong musketry fire. "For a short time the contest was severe," Sherman reported. The Union regiment rallied several times under fire but finally broke and retreated to the road. Now a third regiment, the 69th New York, assaulted. But fire was too heavy, and it likewise fell back in disorder.[14]

The 33rd Virginia found itself too far from home and too exposed and it also withdrew, after having lost one-third of its strength. Shortly after the regiment returned to the defensive line, one of its excited young officers rushed up to Jackson. "General!" he shouted. "The day is going against us!" Jackson gave the man a cold stare and replied, "If you think so, sir, you had better not say anything about it."[15]

Despite the young officer's anxiety, the tide of battle was turning for the Confederates. Although Oliver Howard's brigade finally came forward and regained the guns at the Henry house, it did nothing more.

The deciding blow came from Kirby Smith's Confederate brigade, which arrived on the field and moved to the left, opposite Howard's brigade. Smith was severely wounded as he led his troops into position, and Colonel Arnold Elzey took over command. As soon as the brigade was in position, Elzey ordered his men to charge, driving Howard's dispirited men before them. This sudden, surprise move turned the Federal flank and caused the entire right wing of McDowell's army to stagger.

Shortly thereafter, Jubal A. Early's brigade arrived. Learning from Jeb Stuart that the Federal right was about to break, Early thrust his force to the left of Elzey and ordered it to charge. This was the last straw. Howard wrote afterward, "It was evident that a panic had seized all the troops in sight."[16]

The Federals, some screaming that the enemy was upon them, rushed backward, offering practically no resistance. By the time Early reached Bald Hill, just to the left of Henry House Hill, he could see thousands of Federal troops on the Warrenton Pike heading toward the Stone Bridge and Sudley Springs Ford.

P. G. T. Beauregard now realized that the battle was won. Around 4:00 P.M. he ordered a general charge from the line centered around Jackson's brigade. Wade Hampton's battalion and the 18th Virginia of Cocke's brigade at last swept over the guns. A couple of Beauregard's officers turned some of the guns on the Federal forces, which were now falling into disorder. As the Rebels surged forward, the Union soldiers withdrew, losing discipline and cohesion. Union colonel Erasmus Keyes saw the retreat close up. He wrote later: "As we emerged from the woods [near the Stone Bridge] one glance told the tale; a tale of defeat, and a confused, disorderly and disgraceful retreat. The road was filled with wagons, artillery, retreating cavalry and infantry in one confused mass, each seemingly bent on looking out for number one and letting the rest do the same."[17]

❋   ❋   ❋

P. G. T. BEAUREGARD and Joseph Johnston were so elated by the flight of the Federal army that they did little or nothing to consolidate victory.

Only one senior officer on the field saw that the South had the chance to capture a large portion of the fugitives and to seize Washington itself. Around 5:00 P.M., Jackson, soon to be known universally as Stonewall, the name General Bee had bestowed on him, was having his fractured finger sewed up when he heard from his surgeon that President Jefferson Davis was passing by. The president had arrived a short while before and was touring the battlefield.

"We have whipped them!" Jackson shouted to Davis. "They ran like sheep! Give me 5,000 fresh men, and I will be in Washington City tomorrow morning."

There was no response from the president. Jackson's plea was virtually the only expressed appreciation among Confederate leaders of the great opportunity that lay before them. Since most of the Union soldiers had retreated by the same route they had approached the battlefield— that is, by the roundabout way over Sudley Springs Ford—a brisk move with only a few troops straight up to Centreville would have cut nearly all of them off and forced them to surrender.

The South had been given the chance to end the war with a single additional blow. All that was needed was the resolve—and, above all, the leadership—to bring it off. Every unit best situated to cut the line of retreat should have been put in motion at once, and the senior generals themselves should have taken personal command to inspire and urge pursuit. Certainly the 5,000 Union troops McDowell had left behind to defend Washington would have been in no position to stop an aggressive Confederate attack. They were in a guard position, and with McDowell in retreat they would not have had the leadership they needed to organize and set a defensive position. With the Union's main army totally demoralized, the opportunity was ripe for the South.

Two hours of daylight remained, and that night the moon was nearly full. The Confederates could have pursued the enemy all through the night. But none of the senior leaders thought of consolidating the victory and winning the war—not Davis, not Beauregard, not Johnston. They spent the last hours of daylight touring the battlefield.[18]

Porter Alexander, who witnessed the whole scene firsthand, wrote that the tour doubtless "was an interesting study . . . but it was not war to pause at that moment to consider it. One of the generals—Beauregard, for instance—should have crossed Bull Run at Ball's Ford or Stone Bridge with all the troops in that vicinity, and should have pushed the pursuit all night. Johnston should have galloped rapidly back to Mitchell's Ford and have marched thence on Centreville with Milledge Bonham, James Longstreet, and D. R. Jones, who had not been engaged. No hard fighting would have been needed. A threat upon either flank would doubtless have been sufficient; and, when once a retreat from Centreville was started, even blank volleys fired behind it would have soon converted it into a panic."[19]

General Johnston later excused his failure to capture Washington and end the war. "Our army," he said, "was more disorganized by victory than that of the United States by defeat." Indeed, some troops were disorganized and undisciplined, but this was hardly an adequate justification for his gross failure of command in not attempting to reap the fruits of victory. Large segments of his army were *not* disorganized, and they could have moved with speed.

Beauregard and Johnston did in fact believe they had an opportunity to push on from Manassas, as evidenced by the orders they sent to advance after the fighting ended. The problem was that neither general went in person to deliver the orders and urge the troops forward at this key moment. Consequently, the Confederate advances were called off by rumors—all false and none checked for accuracy—of the presence of enemy troops.

Most notably, Johnston ordered Longstreet and Bonham's brigades to advance on Centreville and intercept the routed enemy forces, but the general did not see to it that they carried out the command. Longstreet's brigade led off, with Bonham following. After the march was well under way, Bonham claimed that his commission was older than Longstreet's, that he was to command the joint operation, and furthermore that his brigade was to go to the front. This bickering consumed precious portions of the last hours of daylight. Just before the advance reached Centreville, Bonham sent forward a squad of cavalry to investigate, a waste of time. If Bonham had merely fired blank cartridges to make a great roar and pretend he was coming, the panic at Centreville would have caused the few defenders to lose heart, and he would have found the village deserted. Instead, Bonham consumed the rest of the daylight in deploying the brigades on both sides of the road.

When the brigades were about to open fire, a major from Johnston's staff arrived with a report that the enemy was moving around to attack the Confederate right, and the two brigades had received orders from headquarters to withdraw behind Bull Run. Longstreet denounced the order as absurd, but Bonham, overwhelmed by responsibility and over-caution, called off the attack. It was discovered afterward that some excitable person had mistaken D. R. Jones's Confederate brigade for a Union force and rushed to report the movement to army headquarters. Without checking whether the report was true or false, one of the staff officers had sent orders, in the names of Beauregard and Johnston, revoking the pursuit.

The truth was that Union forces at this point were not capable of stopping a Confederate advance. At Centreville, McDowell and his generals decided to retreat, but by then it was a moot point. As Captain James B. Fry, a member of McDowell's staff, wrote, "A decision of officers one way or the other was of no moment; the men had already decided for themselves and were streaming away to the rear, in spite of all that could be done. . . . Their tents, provisions, baggage, and letters from home were upon the banks of the Potomac, and no power could have stopped them."[20]

The poet Walt Whitman was in Washington at the time. He wrote: "The defeated troops commenced pouring into Washington over the Long Bridge at daylight on Monday, 22nd . . . baffled, humiliated, panic-struck. Where are the vaunts and the proud boasts with which you went forth? Where are your banners, and your bands of music, and your ropes to bring back your prisoners? . . . The men appear, at first sparsely and shamefaced enough, then thicker, in the streets of Washington. . . . They come along in disorderly mobs, some in squads, stragglers, companies. . . . Amid the deep excitement, crowds and motion, and desperate eagerness, it seems strange to see many, very many, of the soldiers sleeping—in the midst of all, sleeping sound. They drop down anywhere, on the steps of houses, up close by the basements or fences, on the sidewalk, aside on some vacant lot, and deeply sleep. . . . But the hour, the day, the night passed, and whatever returns, an hour, a day, a night like that can never again return. The president, recovering himself, begins that very night—sternly, rapidly sets about the task of reorganizing his forces, and placing himself in positions for future and surer work. If there were nothing else of Abraham Lincoln for history to stamp him with, it is enough to send him with his wreath to the memory of all future time, that he endured that hour, that day, bitterer than gall—indeed a crucifixion day—that it did not conquer him—that he unflinchingly stemmed it, and resolved to lift himself and the Union out of it."[21]

George B. McClellan, who would succeed Winfield Scott and Irvin McDowell as Union commander, confirmed that the Confederacy could easily have won the war in the days after the battle of Manassas. He arrived in Washington on July 26, five days after the battle. He rode around the city. "I found no preparations whatever for defense, not even to the extent of putting the troops in military position," he wrote. "A determined attack would doubtless have carried Arlington Heights [opposite the capital] and placed the city at the mercy of a battery of rifled guns. If the Secessionists attached any value to the possession of Washington, they committed their greatest error in not following up the victory of Bull Run."[22]

McClellan was correct about the error the Confederate leaders made, but he was not as accurate about the key to capturing Washington. Mounting a few guns on Arlington Heights and bombarding the downtown buildings was not the essential matter; severing the capital's rail communications with the North was most critical. Any decent-sized Confederate detachment could have crossed the Potomac at any of the fords above the capital and swept around to the Baltimore and Ohio Railroad, the only railway line leading into Washington from the north. If the B&O was blocked, Washington's food supply would be cut off, and the Lincoln administration would be forced to evacuate the city.

A president who had lost his capital would be unlikely to inspire his people to undertake the onerous and painful task of reunification. Evicting Lincoln from Washington would have devastated the North psychologically at least as much as defeating a Union army on Northern soil would have. Likewise, an insurgent force that had seized the enemy's capital would appear to Britain and France as the most probable winner. This would vastly increase their interest in recognizing the Confederacy and intervening in the war—precisely what President Jefferson Davis hoped would happen.

A strike at the B&O Railroad was not attempted, however. The Confederate leaders in charge in the summer of 1861 were incapable of exploiting the victory at Manassas.[23] The South's first and greatest opportunity to win the Civil War had come and gone. The Confederacy was back where it started—obliged to adopt a strategy to win its independence.

⊰⊱ 2 ⊰⊱

A New Kind of War

HE SOUTH DID not have to win a war against the North, but it had to stop the North from insisting on conquering the South. In the months after the missed opportunity at Manassas, it became clear that the South really had just three possible courses of action—passive defense, active destruction of the enemy army, and invasion of the North.[1] Three leaders emerged to champion each of these strategies. President Jefferson Davis wanted passive defense, Robert E. Lee sought the destruction of the enemy in battle, and Stonewall Jackson pushed for taking the war to the Northern people by invasion.

Each avenue was in conflict with the other two. The problem was vastly complicated by the fact that the Confederacy was saddled with a leader who had already chosen one of these courses—passive defense— and refused to consider the other two objectively. President Davis, the Confederacy's commander in chief, was experienced in military affairs. A graduate of West Point, he had served as a chairman of the military affairs committee of the U.S. Senate and as secretary of war in President Franklin Pierce's cabinet. Davis considered himself a military expert. But in fact, his strategic vision was extremely limited. Thus, instead of bringing together the champions of all three strategies and hammering out the best and most expeditious ways of achieving the South's goals, Davis adhered rigidly to the position he had taken from the beginning: the South was only defending itself against aggression, so all it had to do was fend off Northern stabs until the Northern people became weary or Britain and France intervened.[2]

In other words, Davis was relying on the actions of *others* to determine the fate of the South. Depending on others is a dangerous policy at any time, for it's always uncertain how they may respond to events. It was especially inappropriate in this case because, when Davis guessed wrong about the attitudes of both the Northern people and the political heads of Europe, he had no other policy to fall back on.

The advocates of the other two strategies had to devise alternative ways of getting their ideas accepted. Jackson tried to achieve his aim by enlisting a political ally with the ear of the president. Lee tried to achieve his by deception, telling the president that some patently aggressive move, such as the invasion of Maryland in 1862, was a benign effort to bring the people of that state to the side of the Confederacy, when in fact Lee was seeking to attack and destroy the Union army on Northern soil.

Davis's concept of a passive strategy implied defending every inch of the South that Federal troops violated. A policy of protecting everything, however, ends up protecting nothing. Frederick the Great summarized the principle in the eighteenth century. "Those generals who have had but little experience," he wrote, "attempt to protect every point, while those who are better acquainted with their profession, having only the capital object in view, guard against a decisive blow, and acquiesce in smaller misfortunes to avoid greater."[3] Such a policy was politically difficult for Davis, however, because he was under constant pressure from governors to provide soldiers to defend against local Federal invasions from the sea designed to build bases to tighten the naval blockade. Accordingly, Confederate troops until the very end were spread all over the South, thousands of them standing as useless sentinels along coasts that were never going to be attacked.

Therefore, Davis never fully implemented the one defensive strategy that could have actually stopped the Northern advance—establishing two powerful armies, one northwest of Chattanooga, Tennessee, the other in Virginia, to prevent any deep incursion of Union forces. This strategy would have made the Northern task of conquering the South extraordinarily difficult.

Davis did see the need to protect Virginia and the capital of Richmond. The primary tool for achieving this protection was the Army of Northern Virginia, as the force that had moved up to Fairfax now came to be called. This army became the preeminent military force of the Confederacy, with first call on supplies, equipment, troops, and leaders. Yet even this army was starved for men throughout its history.

Davis tried, as well, to protect Chattanooga and the lateral railways that ran through it and through Atlanta to the south. But he did not see until far too late that possession of Chattanooga—and the approach territory in front of it—was crucial to keeping Northern armies from penetrating deep into the South. He also did not commit sufficient forces early enough to ensure that eastern Tennessee was safe. Finally, a couple of months after the death of Albert Sidney Johnston at the battle of Shiloh in Tennessee on April 6–7, 1862, Davis appointed an old friend, Braxton Bragg, as commander in the West, and kept him long after he had proved himself to be incompetent.[4] As a result, the North cleared Chattanooga of Confederate troops in late 1863, opening the door to the capture of Atlanta and the splitting of the South in two in 1864.[5]

Although Davis continued to advocate a passive defensive strategy, it never had any validity, and the true choices for the South were Lee's policy of seeking to defeat the enemy army in battle and Jackson's proposals to carry the war to the Northern people.

Jackson was not the first general to propose an invasion of the North. Early in October 1861, Generals Joseph Johnston and P. G. T. Beauregard at Fairfax formally asked President Davis to find 20,000 reinforcements to raise the Army of Northern Virginia to 60,000 men and to authorize their invasion of the North. "Success in the neighborhood of Washington," they said, "is success everywhere, and it is upon the northeastern frontier that all the available force of the Confederacy should be concentrated."[6] Davis rejected the proposal, giving as his excuse that he could not denude other places in the Confederacy to concentrate such a large force in northern Virginia. Neither Johnston nor Beauregard pressed the issue further.

It was left to Stonewall Jackson to lay out a true plan for invading the North and to show precisely what it could accomplish. In mid-October 1861, a week after he had been promoted to major general and named to command forces in the Shenandoah Valley, Jackson visited Gustavus W. Smith in his tent at Fairfax. Smith also had just become a major general and been given command of a division in the Army of Northern Virginia. A hero of the Mexican War, Smith was an honors graduate of West Point and had taught at the Military Academy.

Jackson told Smith: "McClellan, with his army of recruits, will not attempt to come out against us this autumn. If we remain inactive they will have greatly the advantage over us next spring. Their raw recruits will have then become an organized army, vastly superior in numbers to our own. We are ready at the present moment for active operations in the field, while they are not. We ought to invade their country now, and not wait for them to make the necessary preparations to invade ours. If the president would reinforce this army by taking troops from other points not threatened, and let us make an active campaign of invasion before winter sets in, McClellan's raw recruits could not stand against us in the field.

"Crossing the upper Potomac, occupying Baltimore, and taking possession of Maryland, we could cut off the communications of Washington, force the Federal government to abandon the capital, beat McClellan's army if it came out against us in the open country, destroy industrial establishments wherever we found them, break up the lines of interior commercial intercourse, close the coal mines, seize and, if necessary, destroy the manufactories and commerce of Philadelphia, and of other large cities within our reach; take and hold the narrow neck of country between Pittsburgh and Lake Erie; subsist mainly on the country we traverse, and making unrelenting war amidst their homes, force the people of the North to understand what it will cost them to hold the South in the Union at the bayonet's point."

Jackson asked Smith to use his influence with Johnston and Beauregard in favor of immediate aggressive operations. Smith responded that he had been present at a meeting at Fairfax a couple of weeks previously when the two generals had proposed an invasion of the North, and the president had rejected it. There was nothing he could do, Smith said.

Jackson rose from the ground where he had been sitting, shook Smith's hand warmly, and said, "I am sorry, very sorry." Without another word, he went to his horse and rode sadly away.[7]

As events were to show, Jackson did not give up on an invasion, and neither did Johnston.

※   ※   ※

SHORTLY AFTER THE battle of Manassas, Jefferson Davis sent Robert E. Lee to West Virginia to recover that rebellious antislavery region for the Confederacy. But the Confederate forces were commanded by two former Virginia governors who spent more time trying to one-up each other than fighting Yankees, and Lee was unable to achieve any gains.

Davis transferred Lee to the south Atlantic coast to build defenses against Union naval landings. In the spring of 1862, Davis brought Lee back to Richmond to serve as his military adviser, and on June 1, 1862, he appointed Lee commander of the Army of Northern Virginia.

For Lee, destruction of the enemy's army was always the best strategy, and he devoted his efforts to accomplishing it. Even on the two occasions when he invaded the North, in 1862 and 1863, his aim was not to destroy Northern factories, farms, and railroads, but to attack and defeat the Union field army in battle.

But destruction of a field army was a far more difficult task in the Civil War than it had been for Napoleon Bonaparte half a century previously. Commanders found this out at least as early as the battles of the Seven Days in the summer of 1862. In the Seven Days, the Confederates under Lee lost one-quarter of their entire army because of direct attacks, and they still were unable to destroy the Union army.

The principal culprit was the Minié-ball rifle. Invented in 1849 by a French officer, Claude-Etienne Minié, it had an effective range of 400 yards, four times that of the smoothbore musket, and was lethal and somewhat controllable out to a thousand yards.[8] As a result, a defending force could fire three or four times as many shots at an advancing enemy with the rifle as it had been able to with the musket. Since a musket had an effective range of only 100 yards, and reloading it took twenty or thirty seconds, defenders could usually get off only two or three shots before the attacking enemy was upon them.[9]

By quadrupling the killing zone through which soldiers had to march, the Minié-ball rifle made orthodox methods of battle so costly as almost to guarantee failure, except in cases of overwhelming force. It was the most deadly weapon ever wielded by infantry up to that time.

Generals were slow to recognize the revolutionary nature of the rifle because they had no method of reaching a decision except to march troops in a line, two men deep, directly on the enemy. Both attackers and defenders also fought standing up, because reloading single-shot rifles was difficult and slow if a soldier lay on the ground. Generals Francis Bartow and Barnard Bee had improvised a sensible response on Matthews Hill in the battle of Manassas—ordering their men to fall on the ground to reload their weapons and to stand up only to fire—but in most subsequent battles generals did not follow their practice. That is because such a response would have required a complete restructuring of infantry tactics, which for over two centuries had called for soldiers to reload while standing in a line. More than anything else, a solution required recognition of the problem—that is, realizing the Minié ball had altered the nature of combat. Jackson most certainly saw it, but it was not grasped by his chief, Robert E. Lee, nor by the vast majority of other commanders on either side. Even after the war, a number of well-known commentators did not recognize the effects of the Minié ball. A British officer, Major General Sir Frederick Maurice, for example, did not mention the Minié-ball rifle in his widely acclaimed 1925 analysis of Lee's leadership.[10]

Soldiers on the attack thus continued to march up on defenders in plain view, making themselves easy targets. Defenders were vulnerable as well, and not only because they were usually standing up. Doctrine rejected allowing defenders in field operations to shield themselves behind fortifications, because they could not form up quickly to counterattack if the enemy was repulsed. Commanders believed that troops behind parapets would be reluctant to abandon them and go over to the attack. In this they adhered to old prejudice, for Napoléon had written in a book, *Le Souper de Beaucaire,* in 1793: "He who remains behind his entrenchments is beaten; experience and theory are one on this point."

Because commanders adhered to the stand-up-and-fight doctrine, defenders were shot down in appalling numbers—so appalling that soldiers began to develop their own defenses. While their commanders continued to position them in the open, the soldiers looked for and used whatever cover they could find—fences, walls, sunken roads, embankments, reverse slopes of hills, even trees. When they knew they were about to be assaulted, they threw up makeshift barriers whenever possible, as, for example, Union soldiers did in the battle of Gaines Mill on June 27, 1862, in the Seven Days. By late 1862, a transformation had taken place. Soldiers by almost universal practice began building field fortifications the moment they got into defensive situations. These included abatis to slow the advance of attackers, parapets to shield defending soldiers, and cleared spaces in front of breastworks to expose attackers to defenders' fire.[11]

Nimble twelve-pounder "Napoleons" (named after Napoléon III, not the emperor) were increasingly rolled up alongside defending troops, where they spewed out twelve pounds of canister—a deadly cloud of metal pellets and fragments—at attacking troops. The terrible trinity of rifles, field fortifications, and canister-firing cannons made attacks all the more deadly.

Such artillery did not provide an answer for attacking troops, however. The defensive power of the Minié-ball rifle proved too much to overcome. When Napoléon Bonaparte encountered a defender who could not be shattered by a frontal attack, he wheeled out smoothbore cannons to within a couple of hundred yards of the enemy—beyond the range of muskets—and knocked a hole in his line with canister. He won most of his later battles in this way. But that approach did not work in the Civil War because the Minié-ball rifle had a range longer than the effective range of canister. When artillery tried to move up close to the enemy, sharpshooters shot down both gunners and horses and usually sent the batteries hurrying to the rear. It was precisely the rifle, plus a bold advance right up to the enemy, that eliminated the batteries of James Ricketts and Charles Griffin that Union General Irvin McDowell had sent forward at Manassas, in hopes of copying Napoléon's method. In the Civil War, the infantry tended to dominate the artillery.

That dominance occurred despite the first widespread employment of rifled cannons, which were long-range, high-velocity pieces used for direct fire.[12] Gunpowder, the only explosive available, was not strong enough to shatter shells into enough pieces to be efficient destroyers of men and matériel, while the inaccurate fuses of the period could not guarantee that shells would explode on target. Rifled cannons also tended to spin out the balls and pellets of canister into a doughnut pattern, with nothing in the center. Finally, rifled pieces buried projectiles in sloping ground. Rifled artillery was useful but not decisive.[13]

Cavalry was not an alternative means of winning battles, either. In Napoléon's day cavalry was sometimes able to crack an opposite infantry line with shock. The best defense against a cavalry charge was to form a square, with musketeers facing outward on four sides. The front rank kneeled with gun butts planted on the ground and bayonets pointed upward. The other ranks pointed bayonets forward. It was difficult for horses to get through such a barrier, and they mostly shied away. Even so, squares sometimes could not be formed in time, and sometimes they were broken. Jeb Stuart's cavalry was able to shake the 11th New York Zouaves at Manassas with a surprise attack. But they did not dislodge the regiment and lost nine men and eighteen horses in short order. In the Civil War, infantry armed with the rifle could usually stop a cavalry charge long before the horses got close enough to break a line. Cavalry in the war lost its decisive role and was reduced to screening advances and retreats, guarding flanks, exploiting routs, reconnoitering, and hit-and-run raids.

✻   ✻   ✻

STONEWALL JACKSON SAW before most anyone else that the Minié-ball rifle had changed the conditions of battle so fundamentally that the traditional approach of assaulting the enemy army head-on—which Robert E. Lee embraced to an extreme degree—was not a rational goal because the cost in lives was too great. He realized that another method had to be devised.

Unusually secretive and reticent in giving his opinions, Jackson never fully explained his thinking in these matters. On just one occasion, early in the war, he had a moment of candor with one of his officers, John D. Imboden, in which he outlined his approach to warfare. Imboden was wise enough to record these words, which Field Marshal Viscount Wolseley, commander of the British army in the 1880s, called "golden sentences" that "comprise some of the most essential of all the principles of war."[14]

Jackson told Imboden: "Always mystify, mislead, and surprise the enemy, if possible; and when you strike and overcome him, never let up in the pursuit so long as your men have strength to follow; for an army routed, if hotly pursued, becomes panic-stricken, and can then be destroyed by half their number. The other rule is, never fight against heavy odds, if by any possible maneuvering you can hurl your own force on only a part, and that the weakest part, of your enemy and crush it. Such tactics will win every time, and a small army may thus destroy a large one in detail, and repeated victory will make it invincible."[15]

In this statement, Jackson was referring to offensive operations in general, but in fact his true concept of war was even more subtle than what he conveyed to Imboden. He made that concept evident not through his words but through his actions—the often unexpected maneuvers he would execute in the two years he led troops against the North. Those maneuvers reflected the solution he devised to the problem posed by the new weapons of the Civil War: force the enemy to attack.

Rather than go on the attack, Jackson saw that he could induce the opposing army to attack at a disadvantage or to move into a position where it was in peril. The surest way to achieve this goal was either to block the enemy's supply line or to stand in the way of where his opponent wanted to go. On occasion he contrived to feign weakness to lure the enemy into moving into vulnerable positions or attacking even if he was not obliged to do so. The enemy, being caught too far forward, could be cut off, or part of his army could be isolated and exposed to a larger Confederate force. An enemy attacking under such circumstances, Jackson was sure, would be soundly repulsed and the soldiers deeply depressed. Then, and only then, could the Confederates attack the enemy and destroy him according to the principles he expressed to Imboden. Jackson's aim, whether offensive or defensive, was always to weaken the enemy so he could then swing around an open flank, drive the demoralized enemy into chaotic retreat, or force the opposing army against some terrain feature such as a mountain range or a river, where it would be compelled to surrender.

This was precisely the approach Jackson would employ to great effect from 1862 until his death in 1863. Of course, up against the competing approaches of his commander in chief, on the one hand, and his commanding general, on the other, Jackson could not persuade the Confederates to wholeheartedly accept his defend-then-attack doctrine, or his idea of taking the war to the Northern people, for that matter. But he was not entirely a prophet without honor. General James Longstreet became a strong advocate of forcing the enemy to attack, though he was reluctant to accept the second part of Jackson's plan, that of attacking the now-demoralized enemy. And Lee on one occasion—at Chancellorsville in May 1863—at last adopted Jackson's defend-then-attack method. But Jackson was mortally wounded in this battle, and Lee at once reverted to direct assaults, a passion that remained until the last. On March 25, 1865, just days before the surrender at Appomattox, Lee lost one-tenth of his tiny remaining army in a hopeless direct assault on Fort Stedman in the Petersburg defenses.

Lee, remarkable leader though he was, never learned how to turn his aggression around and let the Union army beat itself to death against Confederate guns.

       ❋   ❋   ❋

THUS THE STRATEGIES of Jefferson Davis, Robert E. Lee, and Stonewall Jackson collided. That collision of ideas made Davis, Lee, and Jackson the key players in the Civil War. Their actions (and interactions) not only affected the South but in fact determined the course of the entire war—the battles, the campaigns by both North and South, the losses and gains on one side or the other.

How can it be said that just three men, all from the same side, could be the decisive forces in a war that endured for four years, involved hundreds of generals and literally millions of soldiers, and was fought in two separate theaters (East and West)? Because these three were most responsible for devising the strategy for the Confederacy, and that strategy largely dictated the course of the war. From first to last the Union's plan was to overwhelm the smaller Confederate army and force it into submission. (This remained the case even when the Union armies, especially under General George McClellan, proved tentative in executing the plan.) Thus the true strategy of the Civil War arose on the other side, where leaders had to devise ways to neutralize or eliminate the enemy.

To be sure, some Federal commanders, especially Ulysses S. Grant and William Tecumseh Sherman in the West, exhibited considerable innovation.[16] But overall the Confederates could calculate the Union's power in advance and anticipate its lines of thrust. The Civil War unfolded, then, as a series of Confederate responses to the North's predictable strategy. Jefferson Davis directed many of those responses, especially in the West, through his political and command decisions. Meanwhile, Lee and Jackson carried out the campaigns in the East, demonstrating inspired leadership and brilliant field command.

Had these three men resolved the tension among their competing ideas, the Civil War would have proceeded in remarkably different fashion. And had the Confederate leaders adapted to the realities of war in which they were immersed, they might well have ended up victorious.

The Shenandoah Valley Campaign

T HE WINTER OF 1861–62 was a depressing time for the Confederacy. It was bad enough that many of the soldiers were reduced to wearing homespun clothing, that overcoats had disappeared, that brogans had replaced boots, and that there were few tents to go around. Worse, nothing was being done to develop a strategy to win the war. Robert E. Lee was out of the equation, having been sent to Charleston and Savannah to build defenses against sea invasions. Generals Joseph Johnston, P. G. T. Beauregard, and Stonewall Jackson had fallen into silence after having been rebuffed in their proposals to invade the North.

President Jefferson Davis's defensive policy was not working. In the West, Federal forces were gathering for major strikes down the Mississippi and toward Nashville, the Tennessee capital. The U.S. Navy was preparing to seize New Orleans and other ports of the South. Throughout the winter troops poured into Washington at a rate of 40,000 a month. Their avowed intention was to drive to Richmond and end the war. By December the new Union commander, George Brinton McClellan, had turned the Army of the Potomac into a thoroughly efficient, well-armed, and well-equipped machine of 148,000 men.

McClellan was thirty-six years old, a West Pointer (class of 1846) who had performed excellent service in the Mexican War. The dashing, handsome McClellan had received much publicity in driving out several small, ill-led Confederate forces in West Virginia in the spring of 1861. There had been little fighting, but McClellan had issued Bonaparte-like manifestoes that glorified the small actions that did take place. The press was impressed and dubbed him the "Young Napoleon." He caught the fancy of the Northern public and impressed President Abraham Lincoln. On November 1, 1861, Lincoln promoted him to command of all the Union armies, replacing General Winfield Scott.

Although McClellan was an able organizer, he proved hesitant to advance on the Army of Northern Virginia. The reason he cited was the figures provided him by his spy service, operated by a private detective, Allan Pinkerton, who reported that the Rebel army amounted to not less than 150,000 men and was well-drilled and equipped, ably led, and strongly entrenched.[1]

In fact, the Confederate army, under General Joseph E. Johnston, now withdrawn a few miles to Centreville and Bull Run, numbered only 40,000 men, was wretchedly equipped and very poorly drilled, had practically no entrenchments, and, though it numbered able officers among its generals, was still badly commanded, and only formed into divisions as winter came on.

But McClellan would not have attacked Johnston's army even if he had been informed correctly as to its true strength. McClellan, it turned out, had two fatal weaknesses—his compulsion to overestimate the size of the enemy's army and his extreme reluctance to use his army. Confederate Porter Alexander, after seeing firsthand the effects of McClellan's generalship, concluded that he was unfit to command in battle. President Lincoln and his secretary of war, Edwin McMasters Stanton, would ultimately reach a similar conclusion, but in the early days of 1862 "Young Bonaparte" was in command.

McClellan's hesitancy—the constant note in the Northern newspapers was "All quiet along the Potomac"—aroused great restiveness in the people and generated much reproach of the general. But McClellan was oblivious of public disapproval and continued to demand more time to hone his military machine, which by February was approaching 240,000 men.

He was also oblivious of the political implications of his delays. Frustrated with his general's inaction, Lincoln issued a general order for the army to advance on February 22, 1862—George Washington's birthday.

McClellan got a delay, but he didn't see that his inactivity was alienating the one man on whom he depended for his job.

At the end of February McClellan at last submitted a campaign plan to Lincoln. Assuming that Johnston's army at Centreville would continue to block the road to Richmond, he proposed to transfer 150,000 men by sea to the Virginia coast, either Urbanna on Chesapeake Bay, or Fort Monroe, an old post still retained by the U.S. Army on the tip of the peninsula between the York and James rivers southeast of Richmond. The great strength of this plan was that it would exploit Union sea power, both to transfer the army by water and to supply it thenceforth by ships.

But the plan aroused the total opposition of Lincoln, who feared that Johnston might make a dash for Washington. McClellan responded that he would leave plenty of troops to protect the capital, in garrisons in the ring of fortresses he had built around the city, and in field forces nearby. Anyway, he said, once the Union army launched a drive for Richmond, Johnston would be compelled to come at once to defend his capital.

Then, however, a decision by General Johnston and President Davis eliminated the original reason McClellan had given Lincoln for swinging around the Confederate flank by sea. Johnston and Davis realized that their army was exposed so far north, and should be withdrawn while the roads were largely impassable from winter rains, thereby preventing pursuit by the Union army. On March 8, 1862, Johnston began to pull his army back, first to the Rappahannock River, then to the more defensible Rapidan River, north of Orange Court House and Gordonsville.

Johnston's withdrawal required the retreat of Stonewall Jackson, who, with only 4,600 men, had been in Winchester to watch the northern end of the Shenandoah Valley.[2] Johnston told Jackson to defend the Valley as best he could with the troops he had, to use his own judgment, and, if possible, to keep his pursuer, Major General Nathaniel P. Banks, from detaching any large number of his 23,000 troops to reinforce McClellan. Leaving a few hundred horsemen under the reckless and daring commander Turner Ashby to form a cavalry screen, Jackson fell back from Winchester fifteen miles south up the macadamized Valley Pike (now U.S. Route 11) to Strasburg, and then to Mount Jackson, twenty-one miles farther south, hoping to pull the Federal army after him.[3]

Despite the Confederate move, McClellan still pushed hard for an amphibious operation in order to exploit Northern sea power. But he abandoned the idea of landing at Urbanna, because Confederate forces now were close enough to block it. McClellan proposed landing at Fort Monroe, then advancing up the peninsula to Richmond. Secretary of War Stanton wired McClellan on March 13, 1862, that Lincoln "made no objection" so long as enough troops were left to protect Washington. On March 17, McClellan stood on the wharf at Alexandria and saw the first contingent of the army head down the Potomac River and into Chesapeake Bay. McClellan had assembled 113 steamers, 188 schooners, and 88 barges to shuttle the army 10,000 men at a time to Fort Monroe. The transfer was going to take three weeks.

But an ominous change had come over Lincoln and Stanton. Neither was in the least degree knowledgeable about military affairs, but they had decided to take personal control of the war, relegating McClellan to just one of the subordinates who were to do their bidding.

On March 8, Lincoln, without consulting McClellan, had formed the Army of the Potomac into four corps and conceived that the corps commanders—Irvin McDowell, Edwin V. Sumner, Samuel P. Heintzelman, and Erasmus D. Keyes—would take their orders from Stanton, not McClellan. Then on March 12, Lincoln had relieved McClellan as general in chief of all Union armies and left him in command only of the Department of the Potomac, one of seven Lincoln set up in the eastern theater. McClellan learned of the order from friends in Washington who read it in the newspapers. The only excuse Lincoln gave McClellan was that this would allow him to concentrate on the campaign ahead.

It would be scarcely possible to imagine more foolish arrangements than Stanton now made. He scattered Union forces throughout the region and hampered McClellan's efforts on the peninsula. The first blow came when Stanton ruled that 10,000 troops from the garrison at Fort Monroe could not be released without the approval of the fort's commander, which was not forthcoming. Even before the campaign commenced, then, McClellan's force had been reduced by 10,000 men. Much more was to come.

All wars depend upon commanders seizing opportunities. These openings usually appear by planning, much more rarely by fortuitous circumstances. It is even more rare for opportunities to be provided by the enemy. Yet this was to be the case for the Confederacy in the spring of 1862. Even as overwhelming hostile forces were descending on Virginia, the South's unwitting allies were the Union president and secretary of war.

The Confederacy had missed an opening after Manassas. Would the South be in position to seize these new opportunities? President Davis had finally recognized that he needed help badly. He had wired Robert E. Lee in Savannah to return to Richmond as soon as possible, and on March 13, 1862, he had appointed Lee as his adviser, "charged with the conduct of military operations in the armies of the Confederacy." It was not apparent at the time, but the South had at last acquired a pair of consummate leaders. They were occupying decisive positions, Lee at the elbow of President Davis and Jackson pointing the gun barrel of the Shenandoah Valley at Washington.

Unfortunately for the Confederacy, the senior field commander of the army remained the very general who had been unable to organize pursuit of Union fugitives after Manassas and who had never even crossed the Potomac in the weeks of complete Federal impotence following this battle. There would be still more mistakes in these critical early months of 1862.

❊   ❊   ❊

WHEN IT BECAME clear that McClellan was moving a huge army to Fort Monroe, the danger to the Confederacy was at once apparent. If McClellan seized Richmond, he would not only evict the Confederate government but also, because Richmond was a rail hub, essentially end the war in the East. If that happened, the demise of the Confederacy could be measured in months, if not days.

This did not occur at once because McClellan, exercising his usual extreme caution, was stopped by a flimsy line of only 13,000 men that Confederate General John B. Magruder had thrown across the peninsula around Yorktown. Magruder, who was famous in the old army for his amateur theatricals, put on a great show of force that deceived McClellan. On one occasion he marched the same Rebel unit time after time past a spot on a road under Union observation to give the impression of great strength. He sent the unit around through the woods to come past the open point again and again.

McClellan settled down to a long siege of Magruder's line. He was prevented from moving up the wide James River and outflanking the line by the Confederate ironclad *Virginia,* which generally went by its old name, the *Merrimac.* The *Virginia,* based at Norfolk, made all the wooden ships of the U.S. Navy obsolete. Even after a Union armored gunboat, the *Monitor,* had fought the Confederate ship to a draw, the *Virginia* remained a threat.[4]

General Johnston, still in charge of the Army of Northern Virginia, began to move 57,000 men from the Rapidan toward Richmond, leaving only Richard S. Ewell's division of 8,000 men in central Virginia to guard the Orange and Alexandria Railroad and to keep a hand out to Stonewall Jackson in the Valley.

Secretary of War Stanton's wide dispersal of Union forces now became critical. Instead of the huge army of 150,000 men that McClellan had expected, only about 100,000 were at Fort Monroe.

Major General Nathaniel Banks, still in the northern end of the Shenandoah Valley, had moved his 23,000 men into Winchester when Jackson abandoned it. When he realized how tiny Jackson's force was, he got permission to leave 9,000 men in the Valley under James Shields and began to move the remainder of his army east toward Manassas. Banks's orders were to protect Washington and the line of the Potomac while the main army was moving to the peninsula. Then he was to seize Warrenton, on the Orange and Alexandria Railroad southwest of Manassas, and reopen the Manassas Gap Railroad from Manassas to Strasburg to ensure supplies for Union forces that remained in the Valley.

Turner Ashby had been monitoring the withdrawal of Union forces, and on March 21 he reported to Jackson that all but four of Shields's regiments had departed.

Jackson at once saw the opportunity. He was well aware of Lincoln's anxiety about the safety of Washington. He did not yet know the disposition of McClellan's army, but he did know that anything he could do to instill fear in Lincoln's heart might reduce the forces McClellan would

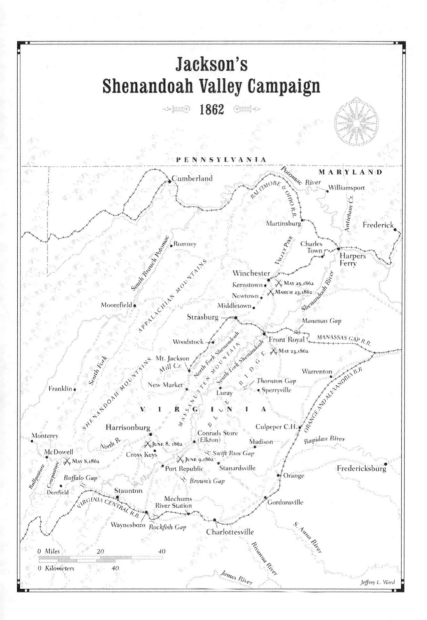

Jackson's
Shenandoah Valley Campaign
1862

PENNSYLVANIA

MARYLAND

Cumberland

Potomac River

Williamsport

BALTIMORE & OHIO R.R.

Antietam Cr.

Martinsburg

Frederick

Romney

Charles Town

Harpers Ferry

South Branch Potomac

VALLEY PIKE

APPALACHIAN MOUNTAINS

Winchester

Kernstown ✕ May 25, 1862

✕ March 23, 1862

Shenandoah River

Moorefield

Newtown

Middletown

Strasburg

Manassas Gap

Woodstock

Front Royal

MANASSAS GAP R.R.

✕ May 23, 1862

South Fork

SHENANDOAH MOUNTAINS

Mt. Jackson
Mill Cr.

North Fork Shenandoah

MASSANUTTEN MOUNTAIN

South Fork Shenandoah

Warrenton

New Market

Thornton Gap

Sperryville

Franklin

Luray

B L U E   R I D G E

ORANGE AND ALEXANDRIA R.R.

V I R G I N I A

Monterey

Harrisonburg

North R.

Culpeper C.H.

McDowell

✕ June 8, 1862

Conrads Store
(Elkton)

✕ May 8, 1862

Cross Keys

Madison

Rapidan River

Ballpasture

Cowpasture

Buffalo Gap

✕ June 9, 1862

Swift Run Gap

Port Republic

Stanardsville

Fredericksburg

Deerfield

Staunton

Brown's Gap

Orange

VIRGINIA CENTRAL R.R.

Mechums
River Station

Gordonsville

S. Anna River

Waynesboro

Rockfish Gap

Charlottesville

Rivanna River

James River

0 Miles   20   40

0 Kilometers   40

Jeffrey L. Ward

have to use against Johnston. Most especially he saw that Banks's departure from the Valley might be reversed by a swift counterstrike. Jackson didn't hesitate for a second. He set his little army in motion. He moved so hard and fast that only 3,000 of his men remained in ranks at 2:00 P.M. on Sunday, March 23, when he came up on Ashby involved in an artillery skirmish with Federals at Kernstown, four miles south of Winchester. Ashby assured Jackson that the force facing them, visible on open ground to the east of the Valley Pike, was only a rear guard. The rest of Shields's force, he said, had already left the Valley.

Jackson told Ashby to demonstrate with his artillery. Meanwhile, he moved the bulk of his troops and guns three miles to his left, or west, up a long, low, wooded hill called Sandy Ridge. He hoped to get on the flank and rear of the Union troops and evict them from their position.

It turned out, however, that Ashby had been wrong about the force opposing him. It was Shields's entire division of 9,000 men, hidden in woods and behind a hill north of Kernstown. Shields had been wounded, but the temporary commander, Colonel Nathan Kimball, sent a brigade onto Sandy Ridge, followed by another, to stop Jackson. Jackson, realizing he was facing a much larger force than his own, ordered his last three regiments to come onto the ridge. But before they could arrive, the Stonewall Brigade commander, Richard B. Garnett, fearing that the thinned Confederate line was about to crack, ordered withdrawal.

Jackson tried to stem the frenzied retreat. But as is usually the case when a force has to withdraw under pressure, the retreat quickly turned into a rout. The remnants of Jackson's little army fled several miles southward, having lost 718 men to the Federals' 590. Nearly one-fourth of his infantry had fallen or been captured.

Jackson relieved Garnett of his command and wanted charges brought against him because he had retreated without orders. His officers and men thought the charge was unjust. But Jackson believed that if Garnett had held a little longer, he could have brought up the reserves and stemmed the Union advance. The army did not pursue a court-martial against Garnett, but thereafter every officer under Jackson knew never to order a retreat until authorized to do so.[5]

General Shields had been stunned by the unexpected Confederate attack. He decided that Jackson would never have attempted it unless he was being greatly reinforced. Fearing for the safety of the Shenandoah Valley, he recalled one of Banks's divisions. On March 24, Lincoln sent Banks and his whole corps back into the Valley. Lincoln also removed Louis Blenker's 10,000-man division, scheduled to go to Fort Monroe, and sent it to strengthen John C. Frémont's army, which was approaching the Valley from the Alleghenies to the west. Thus, this one engagement at Kernstown, though a Confederate defeat, removed 24,000 men from potential use by McClellan.

Worse for the Union general, on April 3 Lincoln ordered Irvin McDowell and his 38,000-man corps, which had stayed behind at Alexandria, to remain near Washington. Secretary Stanton instructed McDowell: "You will consider the national capital as especially under your protection and make no movement throwing your force out of position for the discharge of this primary duty."

Lincoln assured McClellan that McDowell's corps would make its headquarters at Falmouth, next to Fredericksburg, and march overland to Richmond to reinforce him as soon as the threat to the capital disappeared. Although Lincoln released a division under William B. Franklin to go to Fort Monroe, McClellan was paralyzed. He sent wire after wire to Washington trying to get McDowell's corps sent to him.

In his final appeal to Lincoln, on April 5, McClellan complained: "The success of our cause will be imperiled by so greatly reducing my force when it is actually under the fire of the enemy." Lincoln replied: "You now have over 100,000 troops with you. I think you better break the enemy's line at once." McClellan wired back that he actually had only 85,000 men in place at the moment, although his own records showed 108,000. Exasperated, Lincoln responded: "Let me tell you that it is indispensable to *you* that you strike a blow. You must act."

But McClellan did not act. He waited for McDowell. While Jackson was rampaging in the Valley, he was not going to get McDowell. The significance of Kernstown rose higher and higher. Seldom in history has so small a battle had such immense strategic consequences.

* * *

ON APRIL 14, General Johnston came back to Richmond from a visit to Magruder's line at Yorktown. President Davis called a special council of war; joining him and Johnston were Robert E. Lee, Confederate navy secretary George W. Randolph, and Johnston's two senior officers, Gustavus W. Smith and James Longstreet.

Johnston said Magruder's line was untenable. It was subject to artillery fire on the front and to amphibious landings on the flank. Davis read a memorandum from General Smith that Johnston had approved. The memo contained two proposals: (1) immediately withdrawing from the peninsula and from Norfolk, and concentrating all available troops at Richmond against McClellan, or (2) leaving a small force to protect Richmond and striking into Maryland and Pennsylvania with as many troops as could be brought together. Smith wrote that fast-marching Confederates could occupy Philadelphia or New York before McClellan could take Richmond. This Johnston-endorsed plan was premised on the assumption that, because McClellan had locked up the bulk of Union military strength on the peninsula, he would be unable to extricate his army in time to stop the main Confederate army if it marched northward. If Richmond could be shielded, even temporarily, it might be saved, and the war won in the North.

Longstreet later indicated that he had intended to propose an invasion of far less sweeping scope—Magruder to detain McClellan temporarily at Yorktown, and the main army to march on Washington by way of the Shenandoah Valley. This would force McClellan to withdraw to protect Washington, Longstreet believed. But Longstreet did not even raise the idea in the meeting. Thus the senior council was left to consider the two proposals approved by Johnston.

Those proposals were stunning. The Confederacy's senior field commander was endorsing two utterly contradictory ideas—an incredibly far-reaching offensive into the North, or a last-ditch stand before the capital. Neither bore the mark of reality.

It made little sense to give up the peninsula and Norfolk and commit the fate of the Confederacy to a single throw of the dice in front of Richmond. McClellan was so notoriously slow and cautious that much could happen to change the situation if he was fought every step of the way.

The proposal to invade the North was even stranger. Johnston was not advocating a sensible, carefully calculated advance with clear immediate objectives such as isolation of Washington and capture of Baltimore, followed by subsequent goals such as breaking rail communications with the West and the Northeast, all to be determined by the progress of the campaign and the resistance it encountered. Instead he was proposing a full-scale, hell-for-leather offensive all the way to Philadelphia and New York—despite the fact that he had passed on an opportunity even to approach Washington when the Union army was vulnerable after Manassas.

Davis and Lee had every reason to be suspicious of Johnston's invasion idea. And the concept of withdrawing to Richmond and abandoning Norfolk without a fight collapsed from its inherent absurdity. The *Virginia* could not put to sea or ascend the James River because it was unseaworthy and too deep-drafted. If Norfolk was evacuated, the ironclad would have to be scuttled. Without the *Virginia,* the South could not keep Union gunboats from ascending the James.[6]

President Davis ordered the Yorktown line to be held as long as possible, and retreat to Richmond only when it became necessary. In the minds of Davis and Lee doubts mounted as to whether Joe Johnston had the capacity to rise to the challenges of command that shortly were to be demanded of him.

❈   ❈   ❈

STONEWALL JACKSON WITHDREW slowly up the Shenandoah Valley, hoping to find an opportunity to strike at Nathaniel Banks. But the Federal commander was cautious. He pushed his cavalry to Woodstock, about thirty miles south of Winchester, but stopped his infantry at Strasburg, thirteen miles north. Only after McClellan ordered him to clear the Valley did Banks move the infantry to Woodstock.

Jackson was under orders to preserve, if possible, the huge agricultural resources of the Valley and to keep Staunton from capture. Staunton was on the Virginia Central Railroad, which ran from Richmond to a point near Covington. The town was being threatened not only by Banks coming south up the Valley, but also by Frémont marching east over the Alleghenies with more than 15,000 men (plus 10,000 additional men under Louis Blenker on the way). If Frémont and Banks joined at Staunton, the Confederacy would be obliged to abandon the Shenandoah Valley, and the two merged forces then would have an open road to Charlottesville, east of the Blue Ridge Mountains, and Richmond beyond it. Staunton, then, was the key to possession of the Valley.

The total force Jackson had at his disposal was about 17,000 men, fewer than half the Union forces arrayed against him. By recruiting after Kernstown, he had increased his little "Army of the Shenandoah" to 6,000 men. Plus Edward Johnson had 2,800 men facing Frémont at Buffalo Gap, ten miles west of Staunton, and Richard S. Ewell's 8,000-man division was resting not far from the Blue Ridge Mountains near Culpeper. A West Pointer (class of 1840) who turned his head sideways when he talked, the dyspeptic Ewell, like Jackson, believed in speed and traveling light. "The road to glory," he told a subordinate, "cannot be followed by much baggage."[7]

Jackson also had Turner Ashby's Confederate cavalry troopers, who were unrivaled in penetrating the enemy's lines, approaching its camps, observing its columns, deducing what it intended to do, and ascertaining the size and armament of its forces. As one young cavalryman reported, "We thought no more of riding through the enemy's bivouacs than of riding round our fathers' farms." One of Banks's brigadiers remarked: "In reply to some orders I had given, my cavalry commander replied, 'I can't catch them, sir; they leap fences and walls like deer; neither our men nor our horses are so trained.' "[8]

Only in cavalry were the Union forces inferior, however. There were no hardier soldiers in the Union army than those in the Shenandoah Valley, and these men had better equipment and more rifled guns than did Jackson's force. When Banks, under pressure from Washington, finally pushed up the Valley Pike, his troops ousted Jackson from Rudes Hill, an open ridge seventeen miles south of Woodstock and about five miles north of New Market. On April 17 the Federal forces seized New Market, forcing Jackson back to Harrisonburg, twenty-five miles north of Staunton.

Banks's advance to New Market finally opened up the strategic opportunities that Jackson had been waiting for. The Confederate general set out on a forced march to Conrad's Store (now Elkton), which represented the southern anchor to Massanutten Mountain, a huge block set square in the midst of the Shenandoah Valley. Jackson secured Conrad's Store on April 19, leaving a detachment to guard the bridge there over the South Fork of the Shenandoah River and bivouacking in Elk Run Valley at the foot of Swift Run Gap, in the Blue Ridge east of Conrad's Store.

With this move, Jackson had secured Ewell's access to the Valley; Banks could not occupy Swift Run Gap to block Ewell. Jackson sent one of his staff, young Henry Kyd Douglas, on a hurried ride across the Blue Ridge to find Ewell and order him to be ready to move at a moment's notice.[9]

But Jackson—having withdrawn his army from between the two Union forces—seemingly had opened the Valley Pike for Banks to march straight to the Federals' objective, Staunton. In fact, the move to Conrad's Store had slammed the door shut on Banks. Jackson had taken advantage of the Valley's geography. The Shenandoah's bold South Fork ran down the narrow, deeply wooded Luray Valley, which was squeezed between Massanutten and the sharply rising Blue Ridge to the east. Jackson knew that this river was fordable at only a few places and bridged at even fewer. He also knew that only one road crossed the great massif of Massanutten, which ran forty-five miles northeast from Conrad's Store to Strasburg and Front Royal at its northern end. Thus, if Banks marched on Staunton, Jackson would move to Luray, take this one road over Massanutten, and block Banks's supplies and line of retreat at New Market.[10]

In theory, Banks could have marched on Conrad's Store, seized the bridge over the South Fork, corked up Jackson on the other side of the river, and marched on Staunton with his main body. But as Jackson had accurately judged, Banks was far too cautious to attempt such a bold move.[11]

Jackson still had to deal with the threat posed by John C. Frémont, whose advanced brigade, led by Robert H. Milroy, had occupied Monterey and was preparing to continue on the thirty-five miles to Staunton. General Joe Johnston's orders had confined Jackson strictly to a defensive posture, forcing him to remain close to the railway to be ready to move at once to the defense of Richmond if McClellan broke through the lines at Yorktown. But Robert E. Lee, President Davis's new military adviser, saw opportunity in the Shenandoah Valley where General Johnston saw only potential reinforcements for his army. On April 21, Lee wired Jackson: "If you can use General Ewell's division in an attack on Banks, it will prove a great relief to the pressure on Fredericksburg."[12]

This was precisely what Jackson wanted to hear. If he could get 5,000 reinforcements to send to Edward Johnson, he wrote Lee on April 29, Jackson's first preference was to march his army down to Luray, cross at Thornton Gap (present-day U.S. Route 211) to Sperryville on the eastern slope of the Blue Ridge, and from there turn north, leaving the enemy in doubt as to whether he was aiming at Front Royal to the northwest or Warrenton to the northeast, only twenty-five miles from Centreville and the main Federal positions protecting Washington. This march along a single line would immobilize the Washington garrison, keep Irvin McDowell's 38,000 men from marching to aid McClellan, and force Banks into retreat (since his lines of supply would be threatened)—all with practically no casualties.[13]

When Lee could not promise any more troops, Jackson was forced to abandon the concept. But this plan demonstrated in very practical terms how he could implement his admonition to General Imboden always "to mystify, mislead, and surprise the enemy." It revealed both his strategic vision and the depth of his insight.

Jackson turned in a different direction, executing one of the most bewildering moves in military history. He set his little army in motion on April 30, calling on Ewell to move at once to Elk Run Valley and to remain there unless Banks marched on Staunton. Jackson told Turner Ashby merely to "feel out" the Federals toward Harrisonburg. Ashby did not realize that his was a screening move to keep Banks from learning of Jackson's departure or where he was going soon enough to contest it.

Jackson marched south twelve miles to Port Republic, a place just within sight of the enemy's patrols. Then he and his army vanished. The people of Staunton, with Frémont descending on them, were plunged into the direst anxiety. On Saturday, May 3, news came that Jackson's army was crossing the Blue Ridge at Brown's Gap, southeast of Port Republic, and that the Valley had been abandoned to the enemy.

General Banks, meanwhile, was wiring Washington that he was "entirely secure" and that the enemy "is in no condition for offensive movements." He had received a report that Jackson was in retreat, and on April 30 he reported to Secretary Stanton that he was certain "Jackson is bound for Richmond." Banks suggested he leave the Valley, cross the Blue Ridge, and clear the whole country north of Gordonsville.

Jackson's disappearance raised a totally different concern in Stanton's mind. Banks might be right that he was "bound for Richmond," but he also might be heading for Fredericksburg to challenge McDowell. Stanton at once wired McDowell that his first job was to defend Washington, and not to move until Jackson's intentions were better known. Stanton then ordered Banks to send James Shields's division to McDowell as a precaution. Banks's army had been cut in half.

On May 4 Jackson emerged at Mechums River Station on the Virginia Central Railroad, about nine miles west of Charlottesville. But instead of heading to Richmond or Fredericksburg, he loaded his army into train cars and steamed *west* to Staunton. Jackson's maneuvers had landed him in the central position between Frémont on the west and Banks on the north, and neither Union commander could do a thing to help the other.[14] Had Jackson advanced directly from Elk Run Valley on Staunton, he would have advertised his intention, and Frémont would have had long warning of approach. By disappearing from the Valley, Jackson had caused Milroy and Frémont to stop their march on Staunton.

Jackson joined his forces with Edward Johnson's and, with the 200 cadets of Virginia Military Institute as a reserve, marched into the Alleghenies to confront Frémont's advance force. Faced with bad roads and difficult supply, Frémont had spread out his now 25,000-man army in four major segments, thereby negating the numbers advantage he had over Jackson.

On May 8, Jackson confronted 4,000 Union troops at the little village of

McDowell, twenty-seven miles west of Staunton. Though the Union defensive line was strong and Union guns commanded the steep road running down to the bridge, Brigadier Generals Robert C. Schenck and Robert H. Milroy had forced their own troops into a trap. The Union force was in an indefensible position, as the village was in a deep hollow surrounded by high mountains. Jackson sent infantry to occupy Sitlington's Hill, overlooking the village to the east, and made plans to maneuver his main body around McDowell to the north. If he got on the rear and blocked the Federal retreat, the whole force would be compelled to surrender.

Milroy concluded that the Confederates were pulling cannons onto Sitlington's Hill and planned to bombard the village. That was incorrect, since the hill was too steep to drag artillery up. He nonetheless got Schenck's approval to attack it, instigating a perfectly useless battle. The Rebels silhouetted against the skyline suffered the most casualties (498 to 256 Union losses), but Schenck and Milroy at last realized that their position was hopeless and fled westward from McDowell as soon as darkness fell.[15]

Jackson pressed after them, but the Federals delayed him by setting the woods afire and creating a heavy smokescreen. In any case, destroying Frémont's army was not his aim. His purpose was to drive Frémont far back into the mountains so that he could not join Banks or capture Staunton. With Frémont eliminated from the strategic picture, Jackson returned to the Shenandoah Valley to deal with Banks.[16]

❋   ❋   ❋

By May 1862 Jackson's little army in the Shenandoah Valley represented the only ray of hope for the Confederacy. At the end of April, New Orleans, the South's largest city and most important commercial center, fell to the U.S. Navy. Earlier in the month the Confederacy had lost a great battle at Shiloh in Tennessee. The upper portion of the Mississippi, along with Memphis, had fallen to Union troops. Nashville and most of Tennessee had been lost. The senior commander in the West, Albert Sidney Johnston, had been killed at Shiloh. British Field Marshal Viscount Wolseley wrote that by the time the Confederacy lost Memphis and the battle of Shiloh, "the Confederate cause in the West was doomed."[17]

In the East the situation seemed just as bleak. Union General George McClellan had finally advanced to within twenty miles of the Confederate capital. Meanwhile, the South had abandoned Norfolk and scuttled the ironclad *Virginia,* freeing Union gunboats to steam up the James. Only batteries on Drewry's Bluff, six miles below Richmond, barred passage of the gunboats right up to the capital's docks. Military stores were removed from Richmond; the archives were packed and ready to go.

Thus Robert E. Lee wrote Jackson with urgency on May 16: "Whatever movement you make against Banks, do it speedily, and if successful drive him back towards the Potomac, and create the impression, as far as possible, that you design threatening that line."

Jackson was already moving to do just that. Banks had withdrawn to Strasburg, where he posted 7,400 of his men to build strong entrenchments facing the Valley Pike, down which he now feared Ewell or Jackson, or both, might attack. Banks placed 1,000 men at Front Royal, ten miles east of Strasburg on the Manassas Gap Railroad. Jackson had ordered Ewell to follow Banks down the Valley, but now he needed to deal with General Johnston, who wanted Ewell to reinforce the defense of Richmond. Jackson wired the senior commander that Banks was fortifying Strasburg and that the Confederates were moving down the Valley to attack him. Johnston responded that it would be too hazardous to attack in such a situation and that Banks should be left "in his works." Ewell, Johnston said, should come eastward while Jackson remained to observe Banks.

This was not remotely what Jackson had in mind. He telegraphed Lee: "I am of opinion that an attempt should be made to defeat Banks, but under instructions from General Johnston I do not feel at liberty to make an attack. Please answer by telegraph at once." Lee responded authorizing Ewell to remain.

Following Jackson's orders, Ewell sent his Louisiana brigade— commanded by Richard Taylor, the only son of President Zachary Taylor—from Conrad's Store around the base of Massanutten Mountain to New Market.[18] Meanwhile, Jackson's force moved onto the Valley Pike and marched through Harrisonburg north to New Market.

At New Market, the army had no idea where it was going, as Jackson said nothing. On the morning of May 21 the men moved north through New Market expecting that they were marching direct on Strasburg. Just at the edge of the village, however, Jackson quietly turned the head of the column to the right—up the long, sloping road leading over Massanutten Mountain to Luray. The Rebel army had turned completely away from Banks and was marching eastward.

Taylor was as mystified as any of the men in the army. "I began to think," he later wrote, "that Jackson was an unconscious poet, and an ardent lover of nature, who desired to give strangers an opportunity to admire the beauties of his valley."[19]

Jackson's strategy became much clearer to the soldiers when, hours later, the army filed into the Luray Valley and found Ewell's soldiers waiting. In one swift maneuver Jackson had concentrated all of the Confederate troops in his command. With the unified army he turned north toward Front Royal. Now the soldiers realized that the combined force, 17,000 men, was going to fall on the Union flank and rear. The Louisiana Brigade's march to New Market had been intended to deceive Banks; Jackson had also used Turner Ashby to create a cavalry screen.

The deception had worked. In Washington and at Banks's headquarters at Strasburg, complete calm and confidence reigned. On the night of May 22, neither Secretary of War Stanton nor Banks had the faintest suspicion that Jackson had passed beyond Harrisonburg. The Union leaders, including Lincoln, were confident that things were going well. The Confederates had not followed up on the attack on Frémont. McClellan was boasting of imminent success. Lincoln, reassured by Jackson's apparent retreat, had allowed Shields to march his division to join Irvin McDowell. And McDowell, with a portion of his troops, had already crossed the Rappahannock River and was beginning to move beyond Fredericksburg. Lincoln and Stanton, expecting the imminent fall of Richmond, were to leave for Fredericksburg the next day.

Stanton and Banks had done an astonishing job of spreading Union troops in penny packets over the landscape. With Banks at Strasburg were 4,500 infantry and 2,900 cavalry. At the rear base at Winchester were 850 infantry and 600 cavalry. Two companies held Buckton Station on the railway halfway between Strasburg and Front Royal. At Rector-town, east of the Blue Ridge and nineteen miles from Front Royal, were 2,000 infantry and cavalry. At Front Royal, where the North and South Forks of the Shenandoah River came together, Colonel John Reese Kenly had just 1,000 men and two guns to guard the wooden railroad viaduct and bridges that crossed the river.

On the morning of May 23, Jackson, accompanied by the 6th Virginia Cavalry, turned his infantry to the right off the main road and ap-proached Front Royal on a small track winding along the lower slopes of the Blue Ridge. The main body of cavalry crossed a South Fork ford and worked through the forest around the base of Massanutten. During the night Turner Ashby's 7th Virginia Cavalry descended on Buckton to sever Banks from Front Royal. The 2nd Virginia Cavalry under Thomas Taylor Munford destroyed the railway bridges east of Front Royal and set up a blocking force in case Kenly tried to escape in that direction. Meanwhile, small groups of horsemen cut the telegraph between Front Royal and Washington and were ready to block the Union troops posi-tioned at Rectortown.

Thus, around 1:00 P.M., when a long line of Confederate skirmishers broke forward from the forest just south of Front Royal, Colonel Kenly and his garrison were totally isolated. When the Southern cavalry was about to cut him off from the bridges he was valiantly defending, Kenly ordered his troops to withdraw over the North Fork bridge and to fire all three bridges as they went. But the 6th Virginia, accompanied by Jack-son, set off in pursuit. The Confederates encountered Kenly at the ham-let of Cedarville, three miles north of the river, where the Union colonel had pulled up his infantry. With Rebels coming from all sides, the Union force disintegrated as individuals fled in panic. Most were unable to es-cape. The 250 Confederates forced 600 Union soldiers to surrender. In addition, 32 Federals were killed and 122 wounded. Rebel losses were 11 killed and 15 wounded.

Meanwhile, at Buckton, Ashby stormed a log storehouse and drove out the Union garrison. The Federals tried to escape on two trains, but

Ashby's men kept the trains from moving and forced the soldiers to surrender.

Banks, at Strasburg, was strangely unmoved when he heard, at 4:00 P.M., that Front Royal had been attacked. He believed that Stonewall Jackson was still at Harrisonburg, and concluded that the attack was only a cavalry raid. He wired Stanton that it had come from a force that had been "gathering in the mountains."

War is fundamentally a struggle between two intellects rather than a conflict of masses. Jackson, in his surprise descent on Front Royal, was attacking the mind of Nathaniel Banks more than the forces of Colonel Kenly. Jackson had chosen a long and roundabout march rather than hazard a direct attack on Banks at Strasburg. He did this because physical obstacles are inherently less formidable than the hazards of battle. Human resistance is the one great incalculable in warfare. No general can predict human response, and therefore great generals avoid battles whenever they can. In approaching Front Royal, Jackson had taken the line Banks least expected and for which he had prepared the least defense. Banks waited in vain at Strasburg.

Thus Jackson not only avoided a bloody confrontation against Banks "in his works" but also severed Banks's direct rail link with Washington, sealed off his retreat route to the east, placed himself on Banks's flank, and came as close to Winchester, his main rear base, as Banks was himself.

Jackson had affected Banks's ability to make decisions. General George H. Gordon, commanding a brigade, tried to impress on Banks that the attack was coming from Jackson, and that Jackson's aim was to cut off the Union forces at Middletown, five miles north on the Valley Pike. But Banks remained passive, repeating time after time, "I must develop the force of the enemy." Gordon wrote that Banks was "afraid of being thought afraid." This accounts for Banks's outburst when Gordon pleaded with him to move back: "By God, sir, I will not retreat! We have more to fear, sir, from the opinions of our friends than the bayonets of our enemies."[20]

When Banks at last acknowledged his jeopardy at around 10:00 A.M.

on May 24, he realized that his eastern exit was closed and that retreat westward over the Alleghenies was impossible because of bad roads and inadequate supply routes. He therefore ordered an immediate withdrawal to Winchester, abandoning mountains of supplies.

Jackson sent Ewell's main force straight down the direct road from Front Royal to Winchester, and he himself struck toward the Valley Pike at Middletown, in hopes of blocking Banks. Gordon, however, had prudently sent the 29th Pennsylvania Cavalry Regiment east of Middletown to keep the pike open until the army got through. The Union horsemen delayed Turner Ashby's men and the Louisiana brigade long enough for Banks's main body to get north of the village.

The biggest engagement of the day took place at Middletown, after which the Union rear guard fled, mostly by smaller roads to the west. Jackson's infantry moved north on the pike after the fleeing Federals, but the foot cavalrymen were tired after days of hard marching, and the pace was slow. By the time Jackson got to Newtown, six miles farther north of Middletown, he found that Ashby's cavalry advance had melted away. Nearly all of the horsemen had stopped to pillage halted wagons and, especially, to seize horses. Many led captured animals back to their homes, taking one or two days to make the journeys and abandoning the army for this period. The cavalry had to supply their own mounts, and this, to some extent, explains what happened. But the breakdown of discipline largely eliminated Ashby's cavalry as an effective force.

The infantry pursuit also died out from the exhaustion. At last Jackson allowed the soldiers to stop for a couple of hours. Thousands of Rebel soldiers slumped in their tracks and fell asleep in the road.[21]

At four o'clock the next morning, Sunday, May 25, Jackson aroused the sleeping men and set them on the way to seize Winchester before Banks could organize an effective resistance. He had already sent a message to Ewell to be at Winchester at daybreak.

Banks's situation was desperate. He had lost his Front Royal garrison, and about 1,500 men of his rear guard had been scattered at Middletown. He had no more than 7,500 men, and they were discouraged by retreat. Jackson, on the other hand, had about 16,000 men, and they were invigorated by success.

Banks did not help his situation when he failed to deploy more than a skirmish line on the first group of hills south of Winchester. Rather than mount his artillery and main line of defense on this crucial elevation, he set up the Federals' main army on another ridge eight hundred yards farther north.

Still, when the Confederates went on the attack, the Federals resisted stoutly. Ewell fell directly on the Union brigade that faced east of the pike, while two of Jackson's brigades attacked Gordon's brigade on the west. Union rifled guns and infantry rifles, on a ridge about halfway between the Federal line and Jackson's forces, fell heavily on the Confederate troops. That prevented Jackson from engaging the enemy closely on the west and allowed Gordon to extricate two of his regiments and a battery for a movement around the Confederate left flank.

Seeing this, Jackson decided to launch a flank movement of his own. He called up Taylor's Louisiana brigade and William B. Taliaferro's Virginia brigade, which swung to the left and came against Gordon's two regiments, which were firing heavily from behind a stone wall on the hill ahead. The Louisianians and Virginians started up this hill. Halfway up, Taylor gave the order to charge, and the entire line rushed forward at a run. Gordon's men could not stand before this relentless assault, and they tumbled down the back side of the hill. A cavalry charge Gordon ordered failed in the face of the Confederate fire.

The Rebels along the crest of the hill, seeing the Union soldiers reeling from the flank attack, rose as a single body and charged forward, 10,000 men bearing down on the Union lines from two directions and joining in the wild "Rebel yell." The Union soldiers gave way. Jackson rode down the rocky slope waving his cap at the retreating enemy. "Press them forward to the Potomac!" he shouted.

Many of the Union soldiers threw down their arms in Winchester. The others, in increasing chaos, hurried toward the Potomac. Jackson could find no horses to pursue the fleeing enemy. Ashby and the few men he still had with him had gone off without orders to Berryville, thinking the enemy might try to escape through Snicker's Gap in the Blue Ridge. But Ewell's cavalry chief, General George H. Steuart, had 600 horsemen on the right, and Jackson hurried one of his staff,

Lieutenant Alexander S. Pendleton, to bring him up to begin pursuit. When Pendleton found Steuart and gave him the order, Steuart replied archly that he was under Ewell's command and would not move until Ewell said so. Young Pendleton rushed to find Ewell, who authorized Steuart's immediate departure. Yet it took Steuart two hours to get his cavalry under way, and by then the Federals had moved out of range.

Because of a stiff-necked officer and the failure of Turner Ashby, Jackson was unable to destroy Banks's entire army. Jackson had been remiss in allowing Ashby's cavalry virtually a free hand. He moved to correct this mistake in the days and weeks ahead, but on May 25, 1862, a great opportunity was lost because of the want of a few men on horseback.

Even so, the spoils were great. Banks's army suffered losses of 3,000 men, mostly prisoners of war. The Confederates gained over 9,000 rifles, two field guns, many wagons, and rich commissary supplies, especially medicines and bandages, which the South desperately needed.

Most of Banks's survivors found sanctuary at Williamsport, thirty-six miles away on the Potomac, though some got to Harpers Ferry, twenty-five miles away.

      ❖   ❖   ❖

The battle of Winchester created panic in Washington and throughout the North. Now there was no Union force that could stop Jackson from either driving into Maryland and Pennsylvania or swinging around behind Washington to seize Baltimore and sever the capital's rail communications and supply line to the North.

President Lincoln wired Irvin McDowell and ordered him—as well as John C. Frémont in the Alleghenies—to move forthwith to the Shenandoah Valley to cut off Jackson. McDowell was seventy miles from Strasburg as the crow flies, and on the other side of the Blue Ridge. Jackson was on the hard Valley Pike, and it could carry him to safety well before McDowell could arrive.

McDowell protested the order in a wire to Lincoln: "It is impossible that Jackson can have been largely reinforced. He is merely creating a diversion and the surest way to bring him from the lower Valley is for me to move rapidly to Richmond."

Lincoln did not listen. Indeed, his thoughts were turning away from the battle for Richmond. He wired McClellan: "I think the time is near when you must either attack Richmond or give up the job and come to the defense of Washington."[22] Secretary Stanton telegraphed the governors of the Northern states to prepare all of their military forces for a sudden call.[23]

In stopping McDowell for the third time, Jackson had done his damage. His indirect blow had prevented McDowell's corps from combining with McClellan's army in a junction that could have spelled the doom of the Confederacy. Rarely in history has a commander been able to achieve such far-reaching, decisive results with the expenditure of so few resources.

On May 28 Jackson moved the Stonewall Brigade toward Harpers Ferry, followed by other elements of his army. When Jackson appeared on Bolivar Heights above the town, Rufus Saxton's 7,000 Union troops withdrew across the Potomac. But though the Ferry and its magazines might easily have been taken, Jackson made no attempt to follow. His scouts had already informed him that McDowell and Frémont were in motion to cut off his retreat.

Strasburg was the point where the Federal pincers were to come together. This town was the eye of the needle. To get south and into the upper valley, Jackson had to pass through Strasburg. Saxton and Banks, who had partially reorganized his troops after the retreat, were under orders to press down on Jackson from Harpers Ferry and Williamsport. If Jackson could be blocked between Saxton and Banks on the north and between Shields and Frémont at Strasburg, he would be forced to surrender. This is what Lincoln and Stanton were counting on.

On May 29, the main body of Jackson's army marched back to Winchester. But the Stonewall Brigade, along with Jackson himself, remained on the heights above Harpers Ferry.

Jackson seemed blissfully unconcerned about the armies approaching him. He knew something that Abraham Lincoln and Edwin Stanton didn't know. In warfare, uncertainty and apprehension cause commanders to become timid and hesitant. The ordinary general sees a trap waiting to be sprung on the other side of every hill. When several armies are converging on a single point and are separated by a distance or by the enemy, when communication is inexact and difficult, and each general is ignorant of his colleague's movements, ordinary generals become even more hesitant and cautious. These fears were doubled and redoubled in the minds of the Union commanders, for they were trying to cage a general who already had shown astonishing speed and audacity.[24]

*   *   *

WHILE JACKSON PREPARED to deal with the armies converging on him in the Shenandoah Valley, Joseph E. Johnston was sending his army into an attack on McClellan immediately east of Richmond. The two-day battle, May 31 to June 1, known as Seven Pines in the South and Fair Oaks in the North, was mismanaged and confused from start to finish. Johnston had planned a strike by James Longstreet's division around the right, or northern, flank of the two Union corps on the south bank of the Chickahominy River. But Longstreet came up behind D. H. Hill's division on Williamsburg Road (present-day U.S. Route 60) a couple of miles south, achieving nothing but to block other Confederate forces trying to get to the battle. Hill crashed headlong into the defending Federal troops on Williamsburg Road and drove them back, but also achieved nothing.

The Confederates suffered 6,100 killed, wounded, and captured, as against 5,700 Union casualties. In all, the Confederates got about 24,000 men into the fight, but did not engage 16,000 men who were on the field. The Federals, on the other hand, committed all 36,000 of their men, except for a single brigade.

Johnston himself sustained two wounds that were not fatal but were to put him out of action for six months. At the end of the engagement, President Davis realized that Johnston was an inadequate commander in an offensive battle. Though reluctant to lose his right-hand man, he appointed Robert E. Lee commander of the Army of Northern Virginia. Lee thus became the army's third and final chieftain.

Not long afterward, Porter Alexander was riding with Joseph Christmas Ives, who had served briefly on Lee's staff and now worked for President Davis. Alexander asked Ives whether Lee had the audacity to meet the enormous odds the Federals were going to bring against the Confederacy. Ives replied: "Alexander, if there is one man in either army, Federal or Confederate, who is head and shoulders far above every other one in either army in audacity, that man is General Lee, and you will very soon have lived to see it. Lee is audacity personified. His name is audacity."[25]

Ives and Alexander didn't know it yet, but Lee's audacity was going to be manifested in a profound urge to give battle to the enemy in the most direct and challenging manner possible. He believed he could settle the war by aggressive offensives on the battlefield.

Two days before Lee was handed command of the Army of Northern Virginia, another general was proposing an entirely opposite strategy. Far from wanting to collide frontally with the enemy in battle, Stonewall Jackson wanted to *avoid* Union strength—which was expressed most forcefully in its armies in the field—and to strike at Union *weakness,* which was the Northern population, industry, railroads, and farms.

Thus, with President Davis's simplistic plan of resisting Federal incursions into the South having fallen into hopeless ruin, the two remaining views of how the war could be won were about to come into dramatic collision.

❉   ❉   ❉

ON THE MORNING of May 30, 1862, the Stonewall Brigade was making much sound and fury around Harpers Ferry. But it was all sham. With no railway between Winchester and Strasburg, Jackson had pressed his quartermasters to load into wagons all the supplies and arms that had fallen to the Confederacy and haul the spoils of war up the Valley Pike by horses and mules.

Jackson was on the heights above Harpers Ferry to watch an artillery duel and some lively skirmishing. It began to rain, and Jackson got under a tree for shelter and promptly fell asleep. When he awoke, his old friend Colonel Alexander R. Boteler, a former congressman, was sketching the general. Jackson looked at the work, remarked how poorly he had done in drawing at West Point, and said, "Colonel, I have some harder work than this for you to do."

Jackson asked Boteler to go to Richmond and present a proposition to President Davis and to General Lee. Jackson wanted as many reinforcements "as can be spared" to execute his plan, which was essentially a reiteration of the proposal to invade the North that he had made to General Gustavus W. Smith the previous October. If his command was increased to a total of 40,000 men, Jackson told Boteler, he would cross into Maryland, "raise the siege of Richmond and transfer this campaign from the banks of the Potomac to those of the Susquehanna."[26]

This strategic concept was even more possible now than it had been in the fall of 1861, because McClellan had isolated on the peninsula the largest Union army—the only force that could defeat Jackson. Even if Lincoln ordered it to move at once, it would take two weeks at least, and probably longer, to assemble a superior army in Maryland.

Although there were more than 60,000 troops arrayed against him at the moment, they were still scattered. Shields was approaching Front Royal, and Frémont was only a few miles west of Strasburg. So Jackson expected to slip through the trap and withdraw south far up into the Valley. From there, if he got the reinforcements he was asking Boteler to secure, he could burst through or go around[27] any Union forces still in the Valley, cross the Potomac, and have almost free rein until McClellan extricated his army from the peninsula and came to confront him.[28] By then, Jackson indicated, it would be too late.

The North could be in an almost impossible dilemma. The Confederates could cut off Washington's rail communications and food supply, seize Baltimore and perhaps other cities, and spread panic. If Washington was isolated, there would be intense pressure to evacuate the government for fear members would be captured. Britain and France would likely recognize the Confederacy.

The dangers to the Confederacy were much less than those the North faced. McClellan would be unlikely to attack the strong entrenchments that Lee was building in front of Richmond, not only because he had shown himself to be extremely hesitant but also because Lincoln was almost certain to order him to come to Washington's defense the moment he felt threatened.

Therefore, a movement by Jackson into the North at the very least would end the siege of Richmond without a single Southern soldier being sacrificed and would throw the North onto the strategic defensive. It might win the war in the space of weeks.

❋   ❋   ❋

JACKSON AND BOTELER got on the train and headed back to Winchester. On the way, the train was flagged down and a courier rushed on board to hand Jackson a message. It stated that Shields had driven the 12th Georgia Regiment out of Front Royal and that Frémont was getting closer to Strasburg. Jackson seemed unperturbed, however. At Winchester, he calmly prepared papers for Boteler, then saw him off to Richmond. The only caution he exhibited was to order back the 2,000 men of the Stonewall Brigade from around Harpers Ferry. By the night of May 31, the Stonewall Brigade, marching hard, had passed through Winchester and approached Newtown, while Jackson's main army made it to Strasburg.

Both Frémont and Shields acted with extreme caution, just as Jackson had expected. On June 1, Ewell easily held the placid Frémont in check at Cedar Creek, six miles west of Strasburg, allowing the Stonewall Brigade to pass through Strasburg around noon. Only when Ewell withdrew, as the brigade receded up the Valley Pike, did Frémont go after the Confederates.

When Shields did not advance on Strasburg from Front Royal, Jackson suspected that the Union general planned to march south up the Luray Valley, cross Massanutten to New Market, and get on the rear of the Confederate army while Frémont pressed it from the north. Even if Jackson could pass New Market before Shields could get there, Shields might still continue on to Conrad's Store, swing around the base of Massanutten, and block the Rebel army just south of Harrisonburg.

Getting behind Jackson was indeed what Shields intended to do. He ordered his cavalry to speed up the Luray Valley to seize the two bridges that crossed the Shenandoah River's South Fork at Luray and the single bridge over South Fork at Conrad's Store. These three bridges had become all the more crucial because heavy rainstorms had swept the Valley, thrown all of the streams into flood, and made it impossible to ford the river or even to bridge it with pontoons. But when the cavalry reached

Luray on June 2, they found both bridges burned. At dawn on June 4, after a forced night march, the horsemen reached Conrad's Store and found that this bridge as well had been burned.

The instant Jackson had surmised on June 1 that Shields might try to move up the Luray Valley, he had dispatched cavalry parties to break the three bridges. They arrived well before the Federal horsemen. Shields did not know of a fourth bridge at Port Republic. This bridge Jackson did not burn, for he had an important use for it. He sent a party of cavalry to guard it.

Meanwhile, Jackson's army retired up the Valley Pike. Frémont's horsemen pressed the rear guards hard but never were able to slow the march.[29] On June 6, however, as the Confederate army swung southeast from Harrisonburg on the road to Port Republic, Turner Ashby, leading the rear guard, was killed in a bloody clash that turned back a Union infantry and cavalry assault. The death of Ashby affected the entire army, especially Jackson. Ashby was already legendary for his reckless courage, fine horsemanship, and skill at handling his troopers. He would be sorely missed.

On June 6, Frémont got a message through to Shields that Jackson was fleeing toward Port Republic in panic and that a movement by Shields on his flank would break his army into fragments. Shields sent Brigadier General Erastus B. Tyler's brigade to help the advance brigade intercept Jackson. With Frémont pressing from the west and Tyler from the east, Shields expected an easy triumph—this despite the lack of aggression Frémont had shown, and despite the fact that the two small brigades totaled only 3,000 men.

Jackson was indeed on his way to Port Republic. Shortly before arriving, he got a bare message from President Davis regretting that he could send Jackson no additional troops for his proposed offensive to the "banks of the Susquehanna." Colonel Boteler's mission had failed. Jackson took the news stoically but did not give up.

*   *   *

WHEN JACKSON ARRIVED at Port Republic with his advance guard on June 7, he found the bridge still in Confederate hands. The village was located just west of where North River and South River came together to form the South Fork. Over South River was a ford. Over the bolder North River was the bridge that Jackson had been so determined to save. Holding this bridge gave him access to both sides of the river, while denying it to the Federals. Jackson now occupied the central position between Shields on one side of the South Fork and Frémont on the other side. Neither Union force could help the other.

Leaving Ewell's division near Cross Keys, five miles northwest of Port Republic, to block Frémont, Jackson chose to strike at Shields's smaller army first. These Federal elements could still block Jackson's escape route to Brown's Gap over the Blue Ridge, could capture his wagon train, only a short distance below Port Republic, and could press on to Waynesboro, sixteen miles south, and there cut the Virginia Central Railroad.[30] When the contest began on the morning of June 8, the Union advance party scattered Jackson's cavalry vedettes, crossed the South River at the ford, captured part of Jackson's staff and nearly captured Jackson, and threatened the army ammunition train parked just west of the village. Barely escaping over the bridge to the hill above the North River, Jackson hurriedly pulled up artillery that pounded the Union horsemen. A quickly organized infantry assault across the bridge then drove the Federals back across the South River, freeing the captives.

The same morning Frémont, exhibiting unexpected activity, came up on Ewell, who was occupying a strong position on wooded hills a mile and a half southeast of Cross Keys. Although Frémont had twice the force of Ewell on the field, his courage vanished when he saw that the Confederates were waiting for him. Of his twenty-four regiments, Frémont sent just five regiments of Louis Blenker's division of Germans and other immigrants—or "Dutchmen," as the Rebels called them—against the Confederate right, where Isaac R. Trimble's brigade of troops from the Deep South were hidden among oak trees on a flat ridge. Trimble's men lay down and allowed the Union troops to advance to within sixty paces in solid, tight lines. Then the Rebels abruptly rose and released a sheet of flame at the unsuspecting soldiers. The "Dutchmen" staggered, tried to rally, received another shattering volley, and retreated in panic.

Trimble moved a regiment through a sheltered ravine on Blenker's left to threaten a flank attack, driving back Blenker's division in confusion for a mile. That was enough for Frémont, who ordered back Robert C. Schenck's brigade, which had been making a modest advance on Ewell's left. The battle of Cross Keys was over. The Federals had lost 684 men, the Confederates 288.

Jackson left two brigades to watch the now cowering Frémont and gathered the rest of his army at Port Republic to dispose of Shields's advance force, which remained on the other side of the South River and was now a couple of miles northeast of Port Republic. He hoped to accomplish this quickly and then turn back and shatter Frémont. Early on June 9, the Confederates crossed the South River ford and advanced to meet the waiting Federals.

General Erastus B. Tyler had posted his force ably behind a sunken road and small stream running to the river. On the Federal left up against the Blue Ridge was an open "coal hearth" where charcoal was prepared on the flat surface of a hill one hundred or so feet above the valley. Here Tyler had placed seven cannons that commanded the field. Jackson sent two Stonewall Brigade regiments and a battery to assault the guns from the Blue Ridge flank. Mountain laurel slowed the men, however, and the Union gunners and guarding infantry, hearing them coming, drove them back with rifle and canister fire.

Jackson sent the rest of the Stonewall Brigade and a Louisiana regiment directly against the Union forces along the stream and sunken road. Firing from their protective cover, the Federals cut up the Confederates, who retreated back in haste. The Union line advanced with shouts, leading two of Ewell's regiments to hurry forward. They drove in the Union flank briefly, but the Union troops pushed them off and were on the verge of breaking the Confederate line.

Jackson saw that the heavy fire coming from the coal hearth was the key to the battle, and ordered the main body of Richard Taylor's Louisiana brigade, just coming up, to assault those guns. Moving by way of a forest path they discovered, the Louisianians crashed without warning onto the guns and seized the battery. Federal infantry rushed forward and drove off Taylor's men, but some of them, aided by two regiments under Ewell that had also come through the forest, stormed the hearth once more, and this time drove off the Federals.

Union General Tyler saw that he must retake the battery or withdraw. Otherwise the guns, in Southern hands, would dominate his entire position. But he sent his reserves forward too late. Ewell's main body had reached the field, and a brigade moved toward the hearth. As the Federal soldiers advanced, they saw their own cannons turned on them—with Ewell himself serving as a gunner.

The Northerners broke for the rear, pursued by Jackson's cavalry, not absent this day as it had been at Winchester. The horsemen pursued Tyler's force for nine miles until they reached Shields, who had marched desperately toward the sound of the guns and formed a line there.

Jackson, realizing he now had no chance of going back and dealing with Frémont, ordered the small force guarding Frémont to retreat to Port Republic, burn the bridge, and join the army on the south bank. Fearing Frémont might try to join Shields, Jackson called off pursuit of Tyler and marched the whole army to safety into the lower cove of Brown's Gap.

The next day, Shields withdrew to Luray, while Frémont, followed by Confederate cavalry, retreated hastily to Harrisonburg and, ultimately, all the way to Middletown, ten miles south of Winchester. Shields blamed his departure on supposed orders to march with Irvin McDowell's corps to Richmond. But McDowell was going nowhere and had authorized Shields to remain if he had a reasonable chance of defeating Jackson.[31]

In this strange and anticlimactic way the Shenandoah Valley campaign came to an end. His army now alone and unmolested, Jackson brought down his men from the Blue Ridge on June 12, pitched camp just below Port Republic, and gave them a much-needed five-day rest.[32]

The battle of Port Republic had been hard. But a sterling victory was not necessary. Jackson had already gained everything that could be expected by drawing most of McDowell's corps so far from Fredericksburg that it could not be a factor in the upcoming battle for Richmond.[33]

❧   ❧   ❧

JACKSON'S HOPES FOR a counteroffensive into the North had received scant attention from the Confederacy's leaders, despite the sincere efforts of Colonel Boteler. The threat posed by McClellan, the unsatisfactory outcome of the battle of Seven Pines, and the change in command after Johnston's wounding had distracted the leadership. But the real reason was that President Davis was opposed to taking the war to the people of the North, while Lee was focused on the defense of Richmond and on his fixed belief that the way to win the war was to defeat the Union army in battle.

Lee was not opposed to a diversionary action by Jackson, so long as it assisted his maturing plans to confront McClellan. He wrote Davis on June 5 that if Jackson could be reinforced, "it would change the character of the war." But this could be done only by convincing the Carolinas and Georgia to dispatch troops, not by sending any men from Richmond. Davis was not prepared to press these states for men, and the matter lapsed.[34]

Lee reinforced these feelings immediately after the battle of Port Republic. He sent Jackson 8,000 men under Brigadier Generals W. H. C. Whiting and Alexander R. Lawton, seeking to convince Lincoln and Stanton that Jackson was about to launch another offensive. But at the same time Lee directed Jackson to make secret preparations to move his whole army, including the reinforcements, to Ashland, just north of Richmond, to assist in the counterstroke he was preparing against McClellan.[35] To keep up the pretense, Lee allowed the reinforcements to leave Richmond openly by train and Union captives about to be paroled to witness the departure and learn the destination.

While obediently preparing to move to Richmond, Jackson tried to get the government to reverse its strategy and to ensure that the decisive battle *not* be fought at Richmond. On June 13 he called in Colonel Boteler again and asked him to take a letter to Lee to explain his plan and to solicit troops. "By that means," he told Boteler, "Richmond can be relieved and the campaign transferred to Pennsylvania."[36]

Jackson told Boteler to inform Lee that if he could get a total of 40,000 soldiers, he would cross east of the Blue Ridge and proceed northward until he found a gap that would put him on the rear of General Nathaniel

Banks's army (and thus also behind John C. Frémont and the divisions of James Shields and James B. Ricketts). Once these forces were disposed of, Jackson wrote, he would invade western Maryland and Pennsylvania.

Jackson did not outline a precise battle plan. But by divulging that he would march east of the Blue Ridge, he showed that he had already thought out a strategy that would lock all opposition in place. The forces protecting Washington could not move for fear he might strike at the capital. The forces in the Valley had to remain in place for fear he would seize the Shenandoah. By alternately threatening Washington and the Valley, Jackson could prevent the juncture of Union forces. He could defeat any single element that might venture against him.

After Jackson had defeated or bypassed the forces guarding the Valley and Washington, he could cross the Potomac River, beyond which there were no substantial Union field forces. The plan resembled the idea he had advanced to Lee on April 29, when he said he could cross over the Blue Ridge at Sperryville, then proceed north, threatening Front Royal in one direction and Warrenton in another.[37]

Boteler rushed off to see Lee. When he arrived, Lee said: "Colonel, don't you think General Jackson had better come down here first and help me to drive these troublesome people away from before Richmond?"[38] The endorsement Lee wrote on Jackson's June 13 letter made no comment on Jackson's plan for an offensive. Boteler was left to tell Jackson of Lee's decision: the pressure on Richmond prevented the detachment of enough troops for an offensive. But Lee made his opposition to the plan clear when he sent Jackson's letter to President Davis. "I think," Lee wrote, "the sooner Jackson can move this way [toward Richmond] the better—The first object now is to defeat McClellan." Davis endorsed the letter back to Lee, saying, "Views concurred in."[39] Jackson's proposal to transform the strategy of the war had received no more than a passing nod from Lee and Davis.

When he got the news, Jackson made no comment. Leaving his cavalry to guard the Shenandoah Valley, he secretly set his army in motion for Richmond.

4

The Seven Days

S TONEWALL JACKSON HAD tried everything within the bounds of military command structure to induce Robert E. Lee and President Jefferson Davis to avoid a battle with McClellan, to transfer the war to Maryland and Pennsylvania, and thereby to pull the Union army away from Richmond and the peninsula. Jackson's latest plea had come on June 13. But Lee and Davis were focused on the enemy in front of them. The battle Jackson did not want to be fought was going to be fought.

This was a decisive turning point in the war—and a fatal one for the South. Lee, as the top commander in the army, convinced President Davis of his strategy, which was to concentrate the Confederacy's power and directly confront the enemy's strength, his field army. Jackson did not give up on his alternative plan, to strike at the North's weakness—its people's resolve to carry on the struggle—but as a loyal soldier, he had accepted in silence the orders of his military superiors and moved his army toward Ashland.

Ever since taking over command of the Army of Northern Virginia on June 1, 1862, Lee had been working out his plans. He saw that one of Mc-Clellan's corps, 30,000 men under Fitz John Porter, was exposed, positioned north and a few miles west of the remainder of the Union army, about 75,000 men located on the south side of the Chickahominy River. McClellan had placed Porter's corps north of the river to reach out to McDowell's corps, which he continued to hope would approach overland from Fredericksburg, and also to protect the Union army's supply base at

White House, twenty miles east on the Pamunkey River. Lee concluded that a superior force could shatter Porter's corps, sweep down the north bank of the Chickahominy, and threaten McClellan's rail connection with his White House river supply port. To defend this railway, Lee reasoned, McClellan would be forced to come out of his defenses south of the river and confront Lee under unfavorable conditions in the open field. Lee felt he had a good chance of defeating McClellan and forcing his army to retreat in panic or surrender.

As late as mid-June, eleven days before the battle was to open, McClellan was relying on White House for his supplies.[1] But he was already in the process of transferring his supply base to the wider, deeper James River, now that the Confederate ironclad *Virginia* had been scuttled.[2]

Consequently, Lee's entire premise—that McClellan would defend White House and the railway to it—was false. Unknown to Lee, McClellan abandoned White House the moment the battles commenced, and transferred his base to Harrison's Landing on the James, the closest point where transports could tie up safely. The landing was about eighteen miles by road southeast of McClellan's main position, around Fair Oaks and Savage Station.

Porter, far from seeking to prevent access to White House, was in fact concerned *only* with protecting his corps's retreat route to the rest of the Union army on the south side of the Chickahominy. When the Confederate army first collided with Porter's defensive line at Mechanicsville and nearby Beaver Dam Creek, the Union corps was facing *west* to guard White House and the railroad leading to it. When Porter retreated eastward during the night, he abandoned defense of White House and the railroad and turned his corps, on McClellan's orders, to face *north* to guard the bridges over the Chickahominy River to the south, over which Porter was to retreat as soon as McClellan completed transfer of his base to Harrison's Landing and got the remainder of the Union army, south of the river, under way toward it. The next day, in the bloodiest of the engagements that would become known as the Seven Days, Lee fought most of the battle aiming his army in the wrong direction.

From a strategic point of view, the Confederates lost the Seven Days campaign on that second day. Lee's mistake gave McClellan an open road

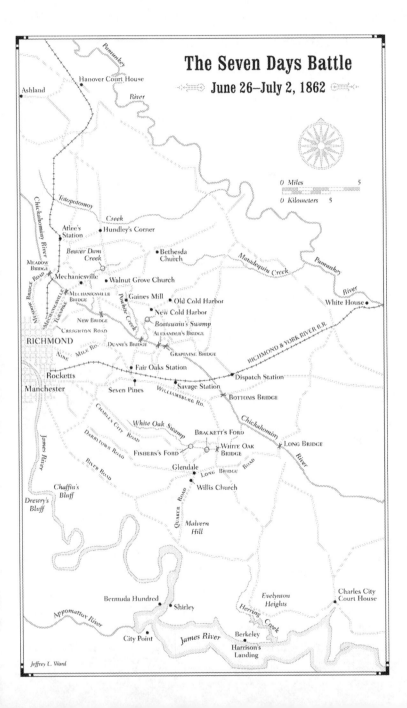

The Seven Days Battle
June 26–July 2, 1862

Pamunkey

Ashland

Hanover Court House

River

Totopotomoy

Creek

Atlee's Station

Hundley's Corner

Chickahominy River

Beaver Dam Creek

Bethesda Church

Matadequin Creek

Pamunkey

MEADOW BRIDGE

Mechanicsville

Walnut Grove Church

MEADOW BRIDGE ROAD

MECHANICSVILLE TPKE.

MECHANICSVILLE BRIDGE

Gaines Mill

Old Cold Harbor

River

White House

NEW BRIDGE

New Cold Harbor

Powhite Creek

CREIGHTON ROAD

Boatswain's Swamp

ALEXANDER'S BRIDGE

RICHMOND

NINE MILE RD.

DUANE'S BRIDGE

GRAPEVINE BRIDGE

RICHMOND & YORK RIVER R.R.

Rocketts

Fair Oaks Station

Dispatch Station

Manchester

Seven Pines

Savage Station

WILLIAMSBURG RD.

BOTTOMS BRIDGE

Chickahominy

White Oak Swamp

BRACKETT'S FORD

CHARLES CITY ROAD

DARBYTOWN ROAD

FISHER'S FORD

WHITE OAK BRIDGE

LONG BRIDGE

River

James River

RIVER ROAD

Glendale

LONG BRIDGE ROAD

Chaffin's Bluff

Willis Church

QUAKER ROAD

Drewry's Bluff

Malvern Hill

Bermuda Hundred

Shirley

Evelynton Heights

Herring Creek

Charles City Court House

Appomattox River

City Point

James River

Berkeley

Harrison's Landing

0 Miles 5
0 Kilometers 5

Jeffrey L. Ward

to Harrison's Landing. When the Union general withdrew, Lee was not convinced that McClellan was heading for the James, even though Porter's dispositions pointed solidly in that direction and Confederate Cavalry Chief Jeb Stuart informed Lee that immense clouds of smoke were rising from White House, certain evidence the McClellan had abandoned it.[3] Lee's hesitation lost twenty-four hours in setting up the pursuit. The Confederates never were able to catch up, though they made a gallant try.

Lee's other mistake came when he did not conclude—from the fact that Porter's defensive line on the second day, June 27, 1862, was facing north, not west—that the entire north bank of the Chickahominy was open. If he had seen this, he could have sent Jackson's corps seven miles straight down to Dispatch Station, the closest point on the railway to White House, and to Bottom's Bridge, a couple of miles below it. Jackson then could have crossed the river and blocked Williamsburg Road (present-day U.S. Route 60). Since the Union army was obliged to retreat for some distance along Williamsburg Road, Jackson could have stopped it on one side while Lee's other forces could have pressed it on the other. If this had been done, McClellan would have been forced to surrender his entire army.

Once more, missteps and missed opportunities kept the Confederacy from turning the tide of the war.

❀   ❀   ❀

THE SEVEN DAYS began on June 26, 1862. Lee had not anticipated a battle on that first day.

He had ordered Jackson's force of 18,000 men to move secretly to Ashland, about sixteen miles north of Richmond. Meanwhile, he had assembled three divisions south of the Chickahominy just opposite Mechanicsville—A. P. Hill's on the west around Meadow Bridge, and D. H. Hill's and James Longstreet's just south of the village.

To guard against an attack straight into Richmond by McClellan's 75,000 men around Fair Oaks, Lee ordered the 28,000 Confederates south of the river to demonstrate as if they were about to attack. General John Magruder, the great Confederate actor who had confused McClellan along the Yorktown line earlier, put on a great theatrical display, making bold gestures and open threats that caused McClellan's commanders south of the Chickahominy to prepare the whole day anxiously for a Confederate attack.

Lee expected Jackson's force to descend on Porter's right flank—a crossroads, Hundley's Corner, about three miles north of Mechanicsville—thereby imperiling Porter's entire position and forcing him to abandon a line of powerful entrenchments he had built along Beaver Dam Creek, a mile east of Mechanicsville. Once Porter realized that Jackson was on his flank, Lee expected, the Union general would have to retreat.

Lee's staff arrangements, however, were entirely inadequate for such a sweeping concentration from various locations on the field of battle. Although his commanders, Jackson especially, did not know the terrain or the roads, he made no provision for communication between the forces; he had sent all of Stuart's cavalry well to the east of Jackson's march. The famous nineteenth-century Prussian army chief of staff Helmuth von Moltke held that uniting two forces on the battlefield is the most brilliant but most difficult feat of generalship. The slightest hesitation by any party may ruin the combination. For Lee and his commanders, none of whom had attempted such an operation, its accomplishment was simply not possible, and it should not have been attempted.[4]

Per Lee's orders, Jackson at 10:30 A.M. on June 26 informed Lawrence O'B. Branch's brigade, located on the Chickahominy about six miles northwest of Mechanicsville, that he had arrived at Merry Oaks, about halfway between Ashland and Hundley's Corner. Branch crossed the river and marched on Mechanicsville but failed to inform A. P. Hill that Jackson was on the way.[5] A. P. Hill, having heard nothing from Branch or Jackson, and fearing that a delay might endanger the whole plan, traversed the Meadow Bridge at 3:00 P.M. and turned his division toward Mechanicsville and Beaver Dam Creek. Meanwhile, D. H. Hill and Longstreet were moving across the river just below Mechanicsville.

A. P. Hill ordered a frontal attack—which Lee approved—against the Union position at Beaver Dam Creek, without artillery preparation and without checking to discover that the Federals had built a wide abatis of logs in front of their position, which completely hobbled the Rebel assault. The result was a terrible slaughter. Hill's division lost nearly 2,000 men in the space of minutes. The Union loss was 360.[6] It was a totally unnecessary and fruitless attack.[7]

Jackson arrived at Hundley's Corner at 4:30 P.M., before Hill's engagement had begun. But he was separated from Hill by a dense forest. Jackson drove off an outpost of Union troops, who reported his presence to Porter. The Confederate general thereupon went into bivouac, having accomplished what he had been ordered to do—place his force on Porter's flank.[8]

During the night Porter fell back on Gaines Mill, a couple of miles to the east, not because of Hill's attack but because Jackson was on his flank. Porter had no doubt about the peril—vast clouds of dust rose above the forests as Jackson arrived, and Union cavalry patrols reported that a huge Confederate force was descending on his north.

※　　※　　※

ON THE MORNING of June 27, Lee, finding Porter gone, assumed that the Union general had moved back to Powhite Creek, a south-flowing stream three and a half miles east of Beaver Dam Creek, and that he was facing westward, protecting the railroad and White House. He ordered A. P. Hill and Longstreet to move directly against this line, and Jackson and D. H. Hill to swing around to Old Cold Harbor, to the northeast of Powhite Creek. Lee reasoned that Porter would be forced to lengthen his line to the east when he saw Jackson and D. H. Hill standing on his path of retreat to Dispatch Station. This would so weaken Porter's position on Powhite Creek, Lee calculated, that A. P. Hill and Longstreet would soon drive the Federals directly past Jackson's position. Stonewall would attack the fleeing Union corps.

But all of Lee's calculations were wrong. Porter was not protecting White House and the railroad. Moreover, he was not lined facing west along Powhite Creek. Instead, his corps was oriented north on a low open plateau behind Boatswain's Swamp, a stream three-quarters of a mile southeast of Powhite Creek. Porter was protecting the Chickahominy crossings. He had left the road to Dispatch Station entirely unguarded. McClellan had ordered Porter to hold this position on Boatswain's Swamp only long enough to cover the withdrawal of his supply train and heavy guns, and to give McClellan time to set up a new supply base at Harrison's Landing on the James.

If Lee had reconnoitered the Federal position, and if he had stopped to reason what this unexpected disposition of Porter's corps signified, he would have seen that a wonderful opportunity had opened to him: McClellan had forgotten to close the back door to his army. Not only was the road to Dispatch Station open, but Jackson's whole force of about 28,000 men (with D. H. Hill) was within easy striking distance of the station and Bottom's Bridge. With Lee in front of him and Jackson behind, McClellan would have had nowhere to go.

But Lee did not see this. He assumed incorrectly that Porter—despite the evidence of his corps facing north—was still protecting White House. He also assumed that the best way to dislodge Porter was to attack him frontally. He ordered A. P. Hill to move forward to do just that.

We have here and in A. P. Hill's mad assault on Beaver Dam Creek the day before two examples of a profound truth about human nature: direct challenge is the normal human response to conflict. We consider anybody who does otherwise to be underhanded, sly, shifty, as not fighting fair. Our language reflects how strongly we feel about this—"stab in the back," "backbiter," and "blindsiding" are all pejorative terms. Reflecting the human tendency to be direct rather than devious or circuitous, most generals throughout history have been unsubtle warriors who attacked the enemy right in front of them.

Yet direct challenge is almost never the way to victory in warfare. The English strategist Basil H. Liddell Hart said that the goal of the great captain is the same as that of Paris in the Trojan War of Greek legend three thousand years ago. Paris avoided any obvious target on the foremost Greek champion Achilles, but instead aimed his arrow at Achilles' only vulnerable point, his heel. Napoléon Bonaparte always attempted to block the enemy's retreat and never made a frontal attack if he could do otherwise. He counted on the menace of a move on the rear, even if it failed, to shake enemy morale and force him into a mistake, which might give Napoléon an opportunity to strike.[9] The great Chinese strategist Sun Tzu wrote around 400 B.C.: "All warfare is based on deception. The way to avoid what is strong is to strike what is weak. Supreme excellence consists in breaking the enemy's resistance without fighting." To achieve this, Sun Tzu recommended that the successful general "march swiftly to places where he is not expected."

English Colonel G. F. R. Henderson eloquently expressed the motivations that led Hill to order the attack on June 26, as well as the dilemmas facing military commanders: "The situation in which Hill found himself, after crossing the river, was an exceedingly severe test of his self-control. His troops had driven in the Federal outposts; infantry, cavalry, and artillery were retiring before his skirmishers. The noise of battle filled the air. From across the Chickahominy thundered the heavy guns, and his regiments were pressing forward with the impetuous ardor of young soldiers. If he yielded to the excitement of the moment, if eagerness for battle overpowered his judgment, if his brain refused to work calmly in the wild tumult of the conflict, he is hardly to be blamed. The patience which is capable of resisting the eagerness of the troops, the imperturbable judgment which, in the heat of action, weighs with deliberation the necessities of the moment, the clear vision which forecasts the result of every movement—these are rare qualities indeed."[10]

At Gaines Mill we see the first full instance of a pattern that Robert E. Lee was to follow throughout the war. When presented, on the offensive, with a defiant enemy in front of him, Lee opted to attack. He did not seek ways around this enemy or hunt for possible mistakes in the enemy's dispositions that would open up other avenues to defeat him. Lee was fixated on direct attack not only at Gaines Mill but also at Frayser's Farm and Malvern Hill in the Seven Days, at Gettysburg, and at the Wilderness. This pattern also operated when Lee was on the defensive. Instead of withdrawing and avoiding battle entirely if a defensive stand could achieve no substantial gains, or instead of looking for defensive positions that offered a chance to maneuver if the attacking enemy was repulsed, Lee pulled up on the most easily held location and challenged his enemy

to attack. In such circumstances, Lee was able to hold his own, but he achieved no strategic gains. Both the battles of Antietam and Fredericksburg were the result of this approach to defensive warfare.

* * *

At 2:30 p.m. on June 27, A. P. Hill sent his infantry to attack the Union forces lined up on the open plateau behind Boatswain's Swamp. Emerging in well-ordered lines from the cover of woods, the Confederates swept down the open slopes, floundered in the swamps, and struggled in the abatis the Federals had hastily laid on the banks of the stream. They drove back the advanced line of Federal infantry along the stream, but that was as far as they got.

"Brigade after brigade," Fitz John Porter reported, "seemed almost to melt away before the concentrated fire of our artillery and infantry; yet others pressed on, followed by supports daring and brave as their predecessors, despite their heavy losses and the disheartening effect of having to clamber over many of their disabled and dead, and to meet their surviving comrades rushing back in great disorder from the deadly contest."[11] The battle raged for nearly two hours, but A. P. Hill's men were repulsed in every instance.

All during this time, Lee had been expecting Porter, whose dispositions were facing largely toward Longstreet on the west and A. P. Hill approaching from the northwest, to shift part of his corps eastward to challenge Jackson and D. H. Hill at Old Cold Harbor. Contrary to Lee's expectations, Porter had left the roads to White House and Dispatch Station almost entirely undefended. Finally, seeing that Porter was not going to move his line, Lee called for Jackson and D. H. Hill to join in a general attack directly against Boatswain's Swamp.[12]

As the day was coming to an end, all of Jackson's forces pushed forward. Union resistance was ferocious, but Confederate power was too great, and at last the soldiers swept in a vast wave onto and over the plateau. The Federal defenders abandoned their last defenses and fled toward the bridges to the south. During the night, after getting across, the retreating Union troops broke the bridges.[13]

Gaines Mill was a terrible bloodbath. Lee lost 8,000 men to Porter's 4,000 and still did not destroy the Union corps. Nevertheless, a remarkable transformation had taken place in the mind of George B. McClellan. Lee's attacks had so stunned him that he moved instantly from an expectation of victory to a fear of defeat. He abandoned any further thought of forcing his way into Richmond by means of his powerful artillery and his much larger army, and now thought only of retreat. This in itself was bizarre. Even after the losses of the first two days of battle, he still had close to 100,000 men, whereas Lee's army had been reduced to about 70,000 men.[14] And McClellan possessed a vast superiority in weaponry.

An extraordinary situation now came to pass. A smaller army set out to pursue a larger army, and the commander of the larger army did not consider any other course of action except flight.

*   *   *

DURING THE MORNING of June 28 ammunition magazines exploded south of the Chickahominy and clouds of dust arose, signaling movement. Lee thus had proof that the Federals were retreating. Since Jeb Stuart's horsemen had found no Federal troops marching toward the highway bridges in the direction of Fort Monroe, and since White Oak Swamp, a wide unbridged marsh and woodland three miles south of the Chickahominy, barred swift passage in that direction, the Union troops' only possible destination was the James River. And their only possible route led along Williamsburg Road to a road about five miles east of Fair Oaks that turned south, crossed White Oak Bridge, joined the Quaker Road at Glendale (Frayser's Farm), then proceeded south over Malvern Hill to River Road, which gave access to Harrison's Landing.

Despite the unequivocal knowledge, Lee did not move until Sunday morning, June 29. When he did finally act, he ordered a complex pursuit plan, even more ambitious than the failed convergence plan to turn Porter out of his Beaver Dam Creek defenses. Lee sent columns down five separate routes. Only one of them, Theophilus H. Holmes's division, which was moving eastward down River Road, had even an outside chance of intercepting the Union army before it reached the James. But Holmes's force was too small, 6,000 men, to challenge McClellan's army. The other four columns moved *behind* the Federals, who had ample time to prepare formidable rear guards.[15]

None of the pursuits succeeded.[16] By Monday morning, June 30, the tail of the Union army had crossed White Oak Bridge, which the Federals then destroyed, while the head of the army was nearing Malvern Hill, on the Quaker Road a mile north of River Road. McClellan ordered William B. Franklin, with 20,000 men, to hold the line at White Oak Bridge until nightfall; Erasmus D. Keyes and Porter to occupy Malvern Hill to shield the army's trains and the landing; and the remainder of the army, about 40,000 men, to defend the area in between, focusing on Glendale.

McClellan, who had been conspicuously absent from the battlefields, now departed once more, leaving no general in charge at Glendale. He rode down to the James, boarded the ironclad *Galena,* and had a good dinner with the skipper and the Comte de Paris, a French nobleman who had joined his suite.

Lee's plan for June 30 was to cut off and destroy the Federal rear around Glendale and around White Oak Bridge. He ordered two of his pursuing generals, Longstreet (with A. P. Hill following) and Benjamin Huger, to continue along the same routes, and instructed Jackson, who had moved eastward in case McClellan turned toward Fort Monroe, to cross at White Oak Bridge. There, Jackson would drive the Union forces toward Huger and Longstreet, who should have been astride the Quaker Road or at least holding the Federals at bay at Glendale. Magruder, who had allowed himself to be stopped by a single Union brigade at Savage Station, three miles east of Seven Pines, no longer served any purpose where he was, so Lee told him to march to Darbytown Road and come up behind Longstreet as a reserve.

If the blows of Jackson, Huger, and Longstreet were coordinated, McClellan could lose his entire rear guard, almost two-thirds of his army. But nothing of the sort took place. Huger failed completely. When he started down Charles City Road, he was seized with fear that enemy troops might attack him from White Oak Swamp, and he sent a brigade to guard that flank, though there was no danger. His other two brigades continued toward Glendale but were stymied by trees the Federals had dropped across the road. Instead of forming teams of soldiers to lift the trees out of the road, Huger cut a new road through the woods. When he finally arrived a mile or so west of Glendale, Huger engaged in a minor artillery duel with a Union division and then stopped, having gained nothing all day.

Jackson also achieved nothing. He faced a force nearly as large as his own at the broken White Oak Bridge. Intense shelling from Federal guns on the ridge just south of the bridge, plus galling fire from Union sharpshooters in the heavy woods along the swamp, prevented Jackson's engineers from repairing the bridge. Frustrated, Jackson made no effort to advance.

If he had reconnoitered, he would have found that the Federals had only about 700 men and one cannon at Brackett's Ford, on a good road a mile upstream from the bridge. Jackson could have broken through at this ford, probably leading to Franklin's defeat, and possibly given the Confederates a great victory at Glendale. Jackson's inaction has never been explained.[17] G. F. R. Henderson concluded that the only fair criticism of Jackson's conduct is that he should have informed Lee of his inability to force passage across the swamp, and should have held three divisions in readiness to march to Glendale by a roundabout route if called upon.[18]

Longstreet reached the approaches to Glendale at noon, with A. P. Hill's division behind. At Frayser's Farm, just south of the crossroads, he encountered a large force of Federals in line of battle. Longstreet waited for Jackson and Huger to drive the Federals toward his position, but the day wore on and nothing happened.

Meanwhile, on River Road, Holmes reached the western face of Malvern Hill. He found only a few Union infantry on it, but Porter had thirty-six guns emplaced. Lee went down from Glendale to look. Seeing that there was a chance to seize Malvern Hill and cut off the retreat of a large part of the Union army, he told Holmes to bring up his whole division and attack, and ordered Magruder to march down and reinforce Holmes.

If Lee had remained to direct the operation, he might have gained Malvern Hill and forced McClellan to attack him in order to get past to Harrison's Landing. But Lee left matters in Holmes's hands and returned to Glendale.[19] Holmes proved incompetent as a combat commander. Though with Magruder he had 20,000 men, far more than Porter had been able to get on the hill, he moved his infantry into the woods and opened fire with six small 3-inch rifled guns. Porter promptly replied with his much bigger guns, reinforced by salvos from Federal gunboats in the James. The barrage produced a lot of noise and some damage, sending green Rebels scampering. Holmes, thoroughly intimidated, withdrew and refused to use Magruder, who returned to the Darbytown Road, having marched all day to no avail.

At Malvern Hill, the cannons were still largely unprotected by infantry. Artillery could do great execution if shielded by infantry, but it could not stand alone against enemy foot soldiers. Riflemen could shoot down gunners and horses, and sweep around the guns and capture them. Thus Lee lost a great opportunity at Malvern Hill.

Back at Glendale, the opposite was the case. A formidable enemy was arrayed in lines of battle and supported by well-protected cannons. Even so, Lee, around 5:00 P.M., realizing that none of the other commanders was going to move, ordered a direct attack straight into the heart of the enemy position. It was the same as he had authorized at Beaver Dam Creek and as he had ordered at Gaines Mill. At Frayser's Farm it was even less logical, for he sent in Longstreet first and then, when he failed, committed A. P. Hill's division.[20]

An attack at this time had little purpose. McClellan was quite obviously moving down the Quaker Road toward Harrison's Landing. Otherwise the Federals would not be defending Malvern Hill so tenaciously. The forces at Glendale and Frayser's Farm would undoubtedly be gone by the morning. The chances of Longstreet and A. P. Hill shattering the Union forces by direct assault were virtually nil. Yet Lee still ordered the attack.

Nowhere did the Confederates gain more than a temporary advantage. Nowhere did they crack the Union line. The two sides stood opposite each other for hours, exchanging one volley after another, while artillery raked the lines of battle. Darkness ended the carnage, but the Federals still held the Quaker Road. By morning they and Franklin's force at the White Oak Bridge were gone. The Confederates lost 3,700 men, all but 220 killed or wounded, the Federals 3,800, half of them captured. It had been a wholly useless battle for the Confederates.

⁕   ⁕   ⁕

DURING THE NIGHT following the battle of Frayser's Farm, the whole Federal army fell back to Malvern Hill. It was a strong position, commanding the country for many miles, and very difficult to assault, especially since Fitz John Porter had lined up all of the army's reserve artillery on it, along with the guns of two corps.

Lee's proper course was not to press on McClellan's heels, which guaranteed another direct assault, this time against the incredibly strong position of Malvern Hill. Instead, Lee should have kept a small covering force on the Quaker Road and marched the bulk of his army straight for Harrison's Landing, on McClellan's rear, in hopes of finding a strong defensive position that could block the Federals, and obligating them to attack the Confederates, not the other way around.

There was such a position, so perfect for its purpose that it might have led to the surrender of the Union army. The place was Evelynton Heights, rising directly north of Harrison's Landing just above the River Road. Every inch of the landing could be hit by guns from Evelynton Heights.

If Lee had occupied Evelynton Heights and brought up all his field pieces, Harrison's Landing would have been untenable. McClellan would have had to abandon Malvern Hill without a fight. To preserve his army he would then have had to assault Evelynton Heights. Such an assault very likely would have failed, because Herring Creek, to the south of the heights, left only a narrow front exposed to attack. Confederate Porter Alexander, Lee's aide Walter H. Taylor, and Jackson's biographer G. F. R. Henderson all thought occupation of Evelynton Heights was by far the best strategy for Lee, even four days later, after much Confederate bloodshed.[21]

But Lee's attention was focused on the retreating Union army. Though he knew the area well—his mother's girlhood home was Shirley Plantation, five miles west of the landing, and he had visited the region often—there is no evidence that Lee considered striking toward Evelynton Heights.

It was obvious to the senior commanders of the Army of Northern Virginia that Lee was planning to attack the Federals on Malvern Hill. The Reverend L. W. Allen, who had been reared in the neighborhood, told D. H. Hill that guns and rifles on the crest and slopes of the hill could sweep the terrain in all directions. Hill tried to talk Lee into canceling the attack or seeking another route to get at the Union army, but Lee ignored him.[22]

The Confederate army arrived in front of Malvern Hill about noon on July 1. Jackson's corps deployed east of the Quaker Road, D. H. Hill's division just west of it, and Huger's on the northwest. Magruder was to

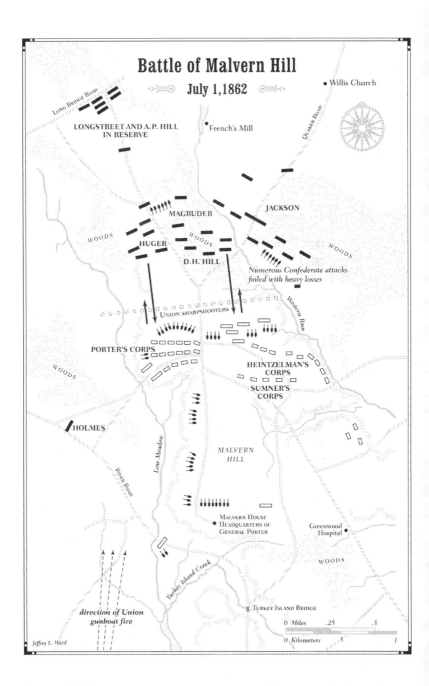

Battle of Malvern Hill
July 1, 1862

• Willis Church

LONG BRIDGE ROAD

QUAKER ROAD

LONGSTREET AND A. P. HILL
IN RESERVE

• French's Mill

MAGRUDER

JACKSON

WOODS

WOODS

HUGER

WOODS

WOODS

D. H. HILL

*Numerous Confederate attacks
failed with heavy losses*

UNION SHARPSHOOTERS

Western Run

PORTER'S CORPS

WOODS

HEINTZELMAN'S
CORPS

SUMNER'S
CORPS

HOLMES

Line Meadow

MALVERN
HILL

River Road

MALVERN HOUSE
HEADQUARTERS OF
GENERAL PORTER

GREENWOOD
Hospital

WOODS

Turkey Island Creek

*direction of Union
gunboat fire*

TURKEY ISLAND BRIDGE

0   Miles   .25   .5

0 Kilometers   .5   1

Jeffrey L. Ward

deploy next to Huger, but he took the wrong road and was hours late coming up. Longstreet's and A. P. Hill's divisions, which had suffered severely at Frayser's Farm, stayed in reserve. Holmes remained on the River Road and did nothing all day.

When Lee arrived on the field, Jackson argued against a direct attack and proposed that the army turn the Federal eastern flank. But Longstreet saw a good position for cannons on an open elevation about a mile west of the Quaker Road and proposed that sixty field guns be mounted there. Lee examined the terrain east of the road where Jackson was emplaced and found a high open field where guns could also be massed. From the two locations, the Confederate cannons could hit the Federal guns from different directions, possibly forcing them to withdraw.

Lee ordered the guns to be brought forward, but the Rebel gunners had tremendous difficulties reaching the two open spaces. The woods through which the batteries had to be forced caused great delays. And when a battery at last emerged at either location and unlimbered for action, it was quickly suppressed by the concentrated fire of fifty or sixty Union guns.

Therefore there was no Confederate artillery barrage against Malvern Hill, while all the Federal guns remained in action.

Lee had made strange and hazardous arrangements for an infantry assault. He designated Lewis A. Armistead, commanding a brigade in Huger's division in the center of the line, to give the signal for the grand assault. Lee's instructions to D. H. Hill and Magruder were as follows: "Batteries have been established to rake the enemy's line. If it [the enemy line] is broken, as is probable, Armistead, who can witness the effect of the fire, has been ordered to charge with a yell. Do the same."

But since the Confederate artillery barrage had failed miserably, Lee decided to cancel the assault and to move—as Jackson had proposed in the beginning—around the east of Malvern Hill with Longstreet's and A. P. Hill's divisions. Unfortunately, he did not convey this order to his staff or get the message to the officers forming up their organizations for the attack.

Sometime around 5:30 P.M. Armistead sent his men a short distance forward to push back Union skirmishers. D. H. Hill, believing he had heard the appointed signal from Armistead, broke out of the woods at the bottom of the hill and advanced with his 10,000 men—against the whole Federal army. Magruder had arrived late and was not ready to join, and none of Huger's regiments took part, either.

The blunder of an unsupported attack—which was even without artillery backing—reaped a terrible toll. A hail of canister fire from cannons hit D. H. Hill's lines from one end to the other. Hundreds fell, other hundreds swarmed back to the woods, but the majority of the men pressed on through the smoke of battle. The Federal infantry waited, lying on the ground, then rose and poured a devastating fire into the Confederates when they reached close range. This stopped the Confederate rush. Here and there small bodies of desperate men continued to follow the colors, but the majority fell to the ground and the whole front exploded with the roar of musketry. As one Federal regiment emptied its cartridge boxes, it was relieved by another regiment. The volume of fire never slackened. Meanwhile, fresh Union batteries unlimbered on the flanks and directed enfilade fire into the stalled Confederate division. The Federals' overwhelming fire at last forced the shattered force back to the woods at the bottom of the hill.

As D. H. Hill was retreating, Huger and Magruder finally came into action on his right. But neither division was formed properly for battle, and the brigades entered piecemeal. The Federal cannons and rifles could concentrate on these smaller units one at a time, and they suffered even more than D. H. Hill's division had. All collapsed and retreated in disorder.

It was a terrible and unnecessary defeat. D. H. Hill wrote: "It was not war, it was murder." Union General Porter said that the Confederate commanders showed "a reckless disregard for life." The cause was the wrong decision by Lee to attack in the first place, but the blunder was vastly aggravated and intensified by Lee's failure to coordinate artillery and infantry and to allow the infantry to attack piecemeal.

The effect on the army was staggering. Isaac Trimble, one of Richard S. Ewell's brigadiers, wrote: "I went off to ask for orders, when I found the whole army in the utmost disorder—thousands of straggling men asking every passerby for their regiments; ambulances, wagons, and artillery obstructing every road, and altogether, in a drenching rain, presenting a scene of the most woeful and disheartening confusion."[23]

The Confederates lost nearly 5,600 killed or wounded, the Federals about 2,000, most left on the field—for the Army of the Potomac withdrew to Harrison's Landing during the night. On the morning of July 2 about a third of the bodies on the field were dead. But a Federal officer who witnessed it said, "Enough were alive and moving to give the field a singular crawling effect."[24]

The Federal soldiers were also demoralized. But the most demoralized of all was General McClellan (who had once again avoided the scene of the fighting and made no provision for a battlefield commander).[25] On the morning of July 1, *before* the battle, he wrote to Washington: "My men are completely exhausted, and I dread the result if we are attacked today by fresh troops. If possible, I shall retire tonight to Harrison's Landing, where the gunboats can render more aid in covering our position. Permit me to urge that not an hour should be lost in sending me fresh troops. More gunboats are much needed. . . . I now pray for time. My men have proved themselves the equal of any troops in the world, but they are worn out. Our losses have been very great. I doubt whether more severe battles have ever been fought. We have failed to win only because overpowered by superior numbers."[26] George McClellan, who had not appeared at a single one of the actions, who had turned and run after his first encounter with a smaller army, could think at the end of the Seven Days only of "more gunboats" and "fresh troops." His had been a pathetic failure of leadership at every level and in every particular.

When the Federals reached Harrison's Landing they collapsed, making no effort to defend the position. The Confederates were little interested in pursuing, and scarcely budged.

※   ※   ※

EARLY ON JULY 3, Jeb Stuart and only a few horsemen managed to evict a small Federal outpost on Evelynton Heights. Stuart then pulled up a single howitzer and began shelling the Union camps at Harrison's Landing. The Federals were initially shocked, but when they sent up a division, they drove off Stuart's small force. What might have happened had the Confederates taken that very hill with a strong occupying force?

The failure to capture Evelynton Heights stands as a symbol of the larger failure of Lee's command during the Seven Days. At Gaines Mill, at Frayser's Farm, and again at Malvern Hill, he opted for head-on battle rather than considering alternative ways of defeating the enemy. These battles also revealed the terrible cost that his method of warfare was going to exact from the Confederacy. Lee had lost 20,135 men in the Seven Days—one-quarter of his entire army. Only 940 of these men were captured or missing; 3,286 were killed and 15,909 wounded. The Federals lost 15,849 men, but only 1,734 were killed and 8,062 wounded, while 6,053 were captured or missing. Since prisoners of war were routinely exchanged on a one-for-one basis, the real Federal loss was only one-half that sustained by the Confederates.

Lee's audacity might win battles, but the cost was going to be more than the South could bear.

5

The Sweep Behind Pope

I N THE SEVEN DAYS two overwhelming facts had become evident. First, the Minié-ball rifle and canister-charged cannons had increased casualties to a point far beyond anyone's imagination or experience. Second, the tactic of lining up forces two men deep and advancing on the enemy was proving to be virtually unworkable, although no one could think of any other method of reaching a decision. Only a single direct assault had succeeded in all the Seven Days—the final charge on Fitz John Porter's Union corps at Boatswain's Swamp. Even in this one successful attack, the casualties had been staggering, in some units 50 percent or more of the members. In all the other attempts, losses had been similarly high, and the attacks had failed.

Robert E. Lee was troubled by the loss of a quarter of his army, so much so that in the next campaign he did not challenge the enemy to a frontal battle. For the first—and, as it would turn out, only—time, he tried to maneuver the Union forces out of Virginia instead of assaulting them head-on. Lee never explained why he resorted to maneuver here, but it most probably was to reduce the toll of dead and wounded. The tactic came out of the eighteenth century, when kings, reluctant to sacrifice their expensive mercenary armies in battle, tried to force their enemies to withdraw by cutting their lines of supply. It often worked, but it almost never resulted in destruction of the enemy force, as Lee found out. After this campaign in the summer of 1862, he realized that the Union army, without the gauge of battle, would remain as strong as ever. Thus he reverted to his familiar pattern of direct confrontation.

Stonewall Jackson likewise was troubled by what he had seen in the Seven Days. The campaign had appeared to the people of the South as deliverance from imminent defeat. Yet the Confederacy was in a worse strategic situation after the Seven Days than it had been before. George McClellan's Army of the Potomac still had 90,000 men and threatened Richmond from Harrison's Landing. And Abraham Lincoln, at last realizing that he and Secretary of War Edwin M. Stanton knew virtually nothing about how to conduct a war, called in a general from the western theater, Henry Wager Halleck, and consolidated all Union forces under his direction. Lincoln also created a new Union "Army of Virginia" out of the various separate commands he and Stanton had spread all over the map.

Realizing that the South was in peril, Jackson made another plea to President Jefferson Davis to take the war to the Northern people. On July 7, 1862, he called on his old friend Alexander Boteler to approach President Davis again. Invading the North, he told Boteler, was the way to bring the Northern people to their senses and to end the war. The South was far too weak in manpower to overwhelm the Union army by sheer force. So the only solution was to avoid the enemy's military strength and to strike at human nature, exploiting most people's willingness to accept peace if faced with the destruction of their property and means of livelihood. McClellan's army was beaten and would pose no danger until reorganized and reinforced. The South should concentrate 60,000 men to march into Maryland and threaten Washington. When Boteler asked Jackson why he didn't present the idea to Lee, Jackson replied that he had done so but Lee had said nothing.

Boteler once more went to Davis. Once more Davis rejected the idea of an invasion. Once more he did not specify his real reason—that he did not want to invade the North because he believed the Northern people would soon tire of the war and quit. Instead he listed a host of excuses: McClellan might be reinforced, McClellan might cut the railroads supplying Richmond from the south, the losses of the Seven Days had been so great that the Army of Northern Virginia could not undertake so dangerous an expedition.[1]

With this final rejection of his invasion plan, Jackson turned to a specific method of overcoming the effects of the Minié-ball rifle. His aim from now on was to defeat the North by destroying one or more of its armies, but Jackson's approach was diametrically opposed to Lee's idea of frontal assaults. Given his reticence, we cannot document the moment when he decided on this new approach, but it almost certainly came after the carnage of the Seven Days and President Davis's rejection of invasion. We know this because Jackson's actions during the campaign against Union General John Pope's Army of Virginia were directed wholly at getting the enemy to attack, not for the Confederates to attack.

Only later would Jackson articulate his new ideas, but even here he did not give a straightforward statement of his thinking. Five months later, at the battle of Fredericksburg, Heros von Borcke, a Prussian officer on Jeb Stuart's staff, wondered aloud whether the Rebels could stop an assault by the vast Federal army arrayed in front of them. Jackson replied: "Major, my men have sometimes failed to take a position, but to defend one, never!" After the battle of Chancellorsville, in the spring of 1863, Jackson said much the same thing to his medical officer, Hunter McGuire: "We sometimes fail to drive them from position, they always fail to drive us."[2] The conclusion we can draw from these statements is that since Rebel soldiers, if occupying a strong position, could always defeat attacks by Union forces, the Confederates should avoid direct attacks themselves, and instead induce the enemy to attack.

But simply stopping Federal attacks was not a recipe for victory, because a repulsed enemy retained his army and could still maneuver or strike again. The only way to victory was to destroy the enemy army. Jackson arrived at a way to do just that. One should place one's troops in a preselected, strong defensive position anchored by cannons, with at least one open flank, and maneuver the enemy into a situation where he wants to attack or is obligated to attack. The attack would inevitably fail, and then one could move swiftly on the open flank of the depressed enemy.[3] Sometimes one could swing around the enemy's flank and throw its force into tumult even without inducing an attack, if the enemy commander had placed his force in a perilous or exposed position.[4]

Once again, however, the reticent Jackson did not expound much on his new theory of warfare. The only record we have of his discussing it comes from Hunter McGuire, after Jackson was wounded at Chancellorsville in May 1863. McGuire wrote that Jackson had told him "he intended, after breaking into [the Federal commander] Hooker's rear, to take and fortify a suitable position, cutting him off from the [Rappahannock] river and so hold him until, between himself and General Lee, the great Federal host should be broken to pieces."[5]

Of course, Jackson's position as a military subordinate to Lee forced him to use persuasion and example to try to convince his commander to follow what was a foreign concept to the bellicose, direct general. This makes it all the harder to document Jackson's efforts, but the recommendations that he made from the Seven Days onward were absolutely consistent in urging his new method of winning the war.

The first testing ground for Jackson's ideas was to come in the campaign against Pope in the summer of 1862. The world watched with breathless excitement the marches, descents on the rear, surprises, battles, and strategies that gave this campaign the tension and drama of high theater. But alongside the spectacle and the glitter, separate from the courage and the sacrifice of heroic and dedicated men on both sides, an intense theoretical struggle was being waged. There was, quite simply, a deep and fundamental divide in the approach of the two men who had emerged as the military leaders of the Confederacy. The outcome of this contest was going to settle the fate of the South.

*   *   *

JOHN POPE, the commander of the newly formed Army of Virginia, had, like Henry Halleck, been pulled in from the western theater. Once united, his army would number more than 60,000 men. President Lincoln ordered Pope to march on Richmond by way of the Orange and Alexandria Railroad, thus approaching the Confederate capital from the west.[6] In the event Pope's army struck at Lee from one direction and McClellan's from another, the Confederacy might die quickly.

But Stonewall Jackson saw a ray of light. If a strong Rebel force could hit Pope before he could consolidate his army, it might throw Pope on the strategic defensive and into retreat before McClellan could come to his assistance. Jackson thought that McClellan was so demoralized he would never strike out on his own. Therefore, it would be little danger to detach a good portion of the Confederate army to go after Pope. Robert E. Lee and President Davis, however, were mesmerized by McClellan's army at Harrison's Landing and extremely reluctant to release troops.

General Pope himself forced Lee's hand. On July 12, 1862, some of his cavalry occupied Culpeper Court House, only twenty-seven miles north of Gordonsville. This move exposed the north-arching "Gordonsville loop" of the Virginia Central Railroad, the only rail line between Richmond and the Shenandoah Valley. The next day, July 13, Lee ordered Jackson, with the Stonewall Division and Richard S. Ewell's division, 12,000 men in all, to move to Gordonsville.

Jackson set about to trap parts of Pope's army and shatter them one at a time.[7] Lee, on the other hand, had no large strategic concept in mind, and shaped his moves to Pope's actions. He expressly stated to Davis in a letter on August 30, 1862, almost at the end of the campaign, that his hope had been only to evict Pope from Virginia: "My desire has been to avoid a general engagement, being the weaker force, and by maneuvering to relieve the portion of the country referred to [central and northern Virginia]."[8]

In the upcoming campaign, Lee tried to avoid a general battle. It was Jackson who brought it on, while Lee, finding battle thrust upon him, was slow to exploit it. Lee and Jackson thus went into the campaign with vastly different goals. Jackson wanted to annihilate Pope's army. Lee wanted to maneuver him back to Washington.

It took Jackson until July 16 to get his leading brigades into Gordonsville. Pope had time to beat him to that critical rail junction, but his cavalry chief, John P. Hatch, waited until he could assemble infantry, artillery, and a wagon train before attacking. He was still some distance away when Jackson arrived and sealed Gordonsville off from anything short of a major assault.

When McClellan learned that Jackson had moved to Gordonsville, his proper move, of course, was to attack the weakened Lee and drive straight into Richmond. But Jackson had been right; McClellan was not going to move on his own volition. And Halleck did not order him to do so.

Halleck was deferring to Lincoln, who had lost confidence in McClellan. After visiting McClellan at Harrison's Landing on July 8, the president determined that the general's army should be brought back to Washington. Thus, when Halleck went to Harrison's Landing on July 24, he told McClellan it was imprudent to keep two Union armies divided, with Lee between them, able to strike at one before the other could come to its aid. McClellan protested, pointing out that his army was in the heart of the Confederacy, only a score of miles from the capital. But Halleck had taken the measure of McClellan. He concluded that "Young Bonaparte" would never go on the offensive, that he didn't understand strategy, and that the army should be pulled back. On August 3 Halleck gave the order to begin withdrawal.

With the decision, Porter Alexander wrote, the Federals "began the evacuation of the only position from which it could have forced the evacuation of Richmond. They were only to find it again after two more years fighting, and the loss of over 100,000 men; and they would find it then only by being defeated upon every other possible line of advance."[9]

Once more the North had failed to press its advantages. The Confederacy would gain still more opportunities because neither of the two new Union generals was adequate, despite having gained some renown in the West. Halleck had graduated third in his class at West Point in 1839 and was called "Old Brains" in the army because he had written scholarly works on war. But he lacked judgment, had little strategic sense, saw dangers at every turn, and was slow, envious of the success of others, and hesitant to make decisions. Pope, meanwhile, was puffed up with pride and confidence. Another West Pointer, he was incautious, ridiculed generals who thought strategy was more important than headlong fighting, and was unable to see danger until it struck him in the face.[10]

But the ardent antislavery Republican had made an excellent impression on Lincoln, Stanton, and the abolitionist congressional Committee on the Conduct of the War. This was at the point that Lincoln became intent on moving the war in a revolutionary new direction. The North had suffered immense losses in the Seven Days and in the battle of Shiloh in Tennessee on April 6–7, 1862, winning the contest but sustaining 13,000 casualties to the South's 10,700. Lincoln feared that the Northern people would not continue making such terrible sacrifices and decided that the only way to victory was to turn the war into a crusade. To do this, he had to promise to free the slaves.

He planned to issue an Emancipation Proclamation but to apply it only to the seceded states, not to slaveholders in the four slave states (Delaware, Maryland, Kentucky, and Missouri) that had remained in the Union. On July 22, 1862, Lincoln read the proposed proclamation to his cabinet. Secretary of State William H. Seward objected. Coming on the heels of the humiliation of the Seven Days, he said, "it may be viewed as our last shriek on the retreat." The proclamation, he argued, should be postponed until the Union attained a military success. Lincoln agreed, pocketed the paper, and waited for a victory.

※   ※   ※

WHEN JACKSON REACHED Gordonsville he was eager to begin aggressive operations, but his force of 12,000 was too small. He appealed to Lee for more troops. Since Pope's menace was growing, Lee ordered A. P. Hill's division, with 12,000 troops, to go to Jackson's support. Hill arrived on July 29.[11]

These moves alarmed Halleck, who was preoccupied with getting McClellan's army away from Harrison's Landing without being attacked in transit by Lee. To distract both Jackson and Lee, Halleck told Pope to make a demonstration toward Gordonsville. Pope's cavalry moved to the Robinson River, a tributary of the Rapidan, only sixteen miles north of Gordonsville.

On August 5, Federal forces reoccupied Malvern Hill and made menacing gestures toward Richmond. Lee was sure McClellan was bluffing, and on August 7 Rebel patrols found them gone. Federal troops began boarding transports for Aquia Creek near Fredericksburg and Alexandria to join Pope. Lee learned this from Captain John S. Mosby, soon to become famous as a partisan raider in northern Virginia. Mosby had just been exchanged after being captured. At Fort Monroe he had heard that

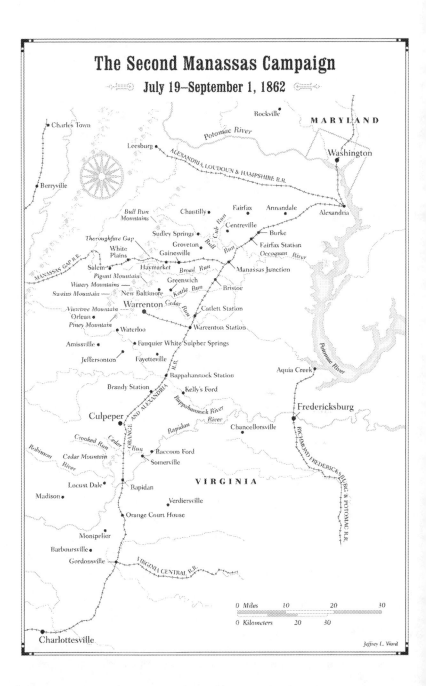

The Second Manassas Campaign
July 19–September 1, 1862

MARYLAND

Rockville

Charles Town

Potomac River

Leesburg

Washington

ALEXANDRIA, LOUDOUN & HAMPSHIRE R.R.

Berryville

Bull Run Mountains

Chantilly

Fairfax

Annandale

Alexandria

Centreville

Sudley Springs

Burke

Thoroughfare Gap

Groveton

Fairfax Station

White Plains

Gainesville

Occoquan River

MANASSAS GAP R.R.

Salem

Haymarket

Broad Run

Manassas Junction

Pignut Mountain

Greenwich

Watery Mountains

Kettle Run

Bristoe

Swains Mountain

New Baltimore

Warrenton

Cedar Run

Catlett Station

Viewtree Mountain

Orlean

Piney Mountain

Waterloo

Warrenton Station

Amissville

Fauquier White Sulpher Springs

Jeffersonton

Fayetteville

Aquia Creek

Rappahannock Station

Potomac River

Brandy Station

Kelly's Ford

ORANGE AND ALEXANDRIA R.R.

Rappahannock River

Fredericksburg

Culpeper

River

Rapidan

Chancellorsville

Crooked Run

Cedar

O Run

Raccoon Ford

RICHMOND, FREDERICKSBURG & POTOMAC R.R.

Robinson River

Cedar Mountain

Somerville

Locust Dale

Rapidan

VIRGINIA

Madison

Verdiersville

Orange Court House

Monipelier

Barboursville

Gordonsville

VIRGINIA CENTRAL R.R.

0 Miles   10   20   30
0 Kilometers   20   30

Charlottesville

Jeffrey L. Ward

a 13,000-man corps under Ambrose E. Burnside, which had been sitting at Newport News, was going not to McClellan but to Aquia Creek. This virtually confirmed that McClellan was abandoning the peninsula.

Pope was confidently developing plans for a victorious campaign in central Virginia that had no basis in reality, given the foe he was facing. He resolved to strike out from Culpeper—where he was trying to assemble his forces—for Charlottesville, forty-five miles southwest. He was sure that Jackson would be forced to pull back to protect this town, site of the University of Virginia, on the Virginia Central Railroad. He wired Halleck: "Within ten days, unless the enemy is heavily reinforced from Richmond, I shall be in possession of Gordonsville and Charlottesville."[12]

Halleck was so excited at the prospect that he ordered Jacob D. Cox, commanding 11,000 Union troops at Lewisburg, to march 130 miles across high mountains and join Pope at Charlottesville. Cox convinced him that such a trek was far too ambitious. "Besides this," he said, showing he knew more about Jackson than the newcomers Halleck and Pope did, "there was the very serious question whether [Pope's] Army of Virginia would be at Charlottesville when I should approach that place." Cox's force took a roundabout journey by boat and rail to Washington, arriving too late to play any role in the campaign.[13]

Jackson had no intention of allowing Pope to capture Charlottesville or of being dragged into a static defense of that town or any other. He had entirely different plans in mind. He pulled his 24,000 men back a few miles south of Gordonsville to the Green Springs area, hoping to lure Pope forward with only a fraction of his command. Warned by Halleck that this might be a ruse, Pope decided to wait until his 50,000 nearest troops had arrived at Culpeper before launching his offensive.

But Pope was having a difficult time gathering these widely separated forces. Informed of this by spies, Jackson saw that if he could seize Culpeper before they arrived, it might become the central position from where he could turn first on one and then on another of the converging Union columns and defeat them.[14]

It was a long shot, but on August 7 Jackson marched his army northward along the Gordonsville-Culpeper road (now U.S. Route 15). Because of traffic tie-ups at the fords over the Rapidan River, however,

Jackson got only eight miles north of Orange Court House on August 8. This delay, plus the fact that his march had been spotted by Union cavalry, gave Pope another day to concentrate, and caused Jackson to wonder whether he had lost his chance.

Though Pope's whole force might reach Culpeper before he could do so, Jackson decided the odds were still in his favor, and launched his army for the town early on August 9. His lead element, Jubal A. Early's brigade of Ewell's division, preceded by a screen of cavalry, crossed the Robinson River just north of the crossroads of Locust Dale, and shortly before noon advanced up a small valley bounded a mile on the east by the rounded ridge of Cedar Mountain, more or less parallel to the road. He was about seven miles south of Culpeper. The land between road and mountain was mostly open meadow or fields planted in Indian corn. On the left, or west, of the road the terrain rose to a forested upland, except for a rectangular wheat field of about forty acres cut out of these woods with one side fronting the road.

Cavalry patrols reported that Federal horsemen in some strength were massed a mile or two north on the banks of Cedar Run. Jackson had already learned from spies that a unit of Nathaniel P. Banks's corps had been the first into Culpeper, and surmised that this force ahead of him was part of the same corps. He was not at all displeased, and had told his medical director that Banks "is always ready to fight, and he generally gets whipped."[15]

There was no sign that Banks had any backup on Cedar Run. Jackson had failed to catch a separated part of Pope's army, but Pope—by sending Banks forward all alone—had conveniently done it for him.

Banks had about 9,000 men and a picket of cavalry. Pope's orders to Banks, delivered by a member of his staff, Robert E. Lee's nephew, Colonel Lewis Marshall, were vague enough to give the impulsive Banks, still smarting from the defeats Jackson had handed him in the Shenandoah Valley, the leverage to attack in hopes of redeeming his reputation.

The closest other elements of Pope's army were James B. Ricketts's division of 9,200 men, still some distance behind Banks; Rufus King's division, on the road from Fredericksburg, and John C. Frémont's old force (now commanded by Franz Sigel upon Frémont's resignation), hurrying from Sperryville. None could get to the field in time to help Banks on August 9.

The battle of Cedar Mountain was a model of the way the Civil War was fought in the summer of 1862. It demonstrated the continued stand-up nature of infantry assaults, the failure of defenders to emplace behind bulletproof barriers or parapets, an atavistic idea that cavalry could turn the scales of battle by charges, the increasingly sophisticated positioning of artillery, and the difficulty of feeding troops into combat unit by unit by an army obliged to approach the battlefield on a single road.

Jubal Early's brigade deployed into the fields to the right of the highway. Beyond Cedar Run, about 1,200 yards north, Federal artillery opened, causing Early to withdraw to the reverse slope of a low ridge running from the road to Cedar Mountain. Jackson ordered all of his artillery, twenty-six guns, to be pushed forward. He posted eight guns in the center of the field, eight on the forward, open slope of Cedar Mountain, and the remainder along the road

This placement allowed the Confederate cannons to dominate the whole field. Any Union advance would be subject to deadly enfilade fire from the elevated position on the mountain. When the Rebel guns opened on the Union batteries, the artillery thundered across the valley for nearly two hours.

Meanwhile, Ewell's division took position on the northern face of Cedar Mountain, while Jackson's division, under Charles S. Winder, deployed in the woods to the left of the road, and the Stonewall Brigade stayed back behind the guns. A. P. Hill, still coming up (Jackson's column on the road was seven miles long), was to form the reserve in the rear of Winder.

The wheat field on the left front was the most vulnerable point for the Confederates. On its northern edge a dense wood provided cover where Federal troops could mass for attack, while to the west stretched a great tract of forest, through which enemy infantry might infiltrate and strike Winder's unprotected flank.

Jackson sent word to Lieutenant Colonel Thomas S. Garnett, commanding the brigade on this flank, to "look well to his left, and to ask his divisional commander for reinforcements." Garnett dispatched two officers back to Winder to get support. But at this juncture General Winder was mortally wounded by a shell. In the ensuing confusion the storm broke on Garnett's front and flank.

Banks, around 5:00 P.M., gave orders for a general attack. Two Union brigades bore down on the Confederate center, defended by Early and William B. Taliaferro's brigade. Though the soldiers advanced bravely and resolutely, they were stopped by Rebel rifles and by the guns, especially those firing from the height of Cedar Mountain. But on the west in the edge of the woods bordering the wheat field—the very point Jackson had seen was vulnerable—only two Confederate regiments, the 1st Virginia and the 42nd Virginia, had been posted, 500 men in all. From the woods on the other side of the wheat field 1,500 Federals emerged, bayonets fixed and colors flying. The Virginians exacted heavy punishment as the Union soldiers marched the 300 yards across the wheat field. But the Federal line was much longer than the Rebel line, and its western wing swung entirely around the 1st Virginia on the extreme left and threw it into panicked retreat. The flight of the 1st Virginia dislodged the 42nd Virginia as well, and the whole Union line descended on the artillery along the road and on the unprotected left side of Jackson's division (now commanded by Taliaferro).

The first warning the Rebels on the road received was a sudden storm of musketry, loud cheers of the Union infantry, and the rush of Confederate fugitives from the forest. Jackson was there, and at once he ordered the cannons to be withdrawn. Not a gun was lost. But the Union infantry, with the Rebel artillery gone, charged forward with reckless courage. Every regimental commander in Garnett's brigade was killed or wounded. Taliaferro's brigade near the highway was driven back, and Early's brigade in the field beyond was broken. Some regiments tried to turn to meet the assault, but others collapsed and fled. Officers rushed into the melee trying to restore order. For a brief time Rebels and Yankees fought in close combat amid the smoke and chaos of battle.

Fortunately for the Confederates, Banks had not provided any support for his right wing. The Stonewall Brigade—five small but sturdy regiments advancing from behind where the guns had been—broke through the rout and opened a heavy fire. The Federals were still in superior numbers but had lost all order. To meet this fresh danger, they halted and drew together.

Jackson sent orders for Ewell and Hill to attack at once. Drawing his sword, he galloped into the midst of his confused and fleeing soldiers, yelling, "Rally men and follow me!" His familiar voice and figure emboldened the men who heard him, and started to suppress the panic. At that moment, General Taliaferro rode up to Jackson and in emphatic tones told him this was no place for the leader of an army. Jackson looked at Taliaferro in surprise, then, saying, "Good, good," turned away to the rear.

But the recovery had occurred. The remnant of the 21st Virginia of Garnett's brigade had gotten itself together, charged forward, and delivered a solid volley into the massed Union soldiers. Other soldiers bore their colors forward and many of the men followed. Taliaferro and Early reformed their brigades and advanced on the Federals from the right.

His front once more established, Jackson turned to the counterstroke he had already ordered. To support the Stonewall Brigade, Jackson sent forward Lawrence O'B. Branch's North Carolina brigade of Hill's division. He sent Edward L. Thomas's Georgia brigade to assist Early, and ordered James J. Archer's Georgia and Tennessee brigade and William Dorsey Pender's North Carolina brigade to swing far around the enemy's right or western flank.

Pressed by the Stonewall Brigade, the Union forces, exhausted and in a desperate situation, retreated back into the woods, across the wheat field, and into the forest beyond, managing to regroup enough to throw two regiments of the advancing Stonewall Brigade into confusion briefly. With Hill's brigades pressing on the western flank and Ewell starting forward next to Cedar Mountain on the east, the Federals fought valiantly but could make no headway.

Banks made a last despairing effort. He ordered two squadrons of the 1st Pennsylvania Cavalry to charge down the eastern edge of the wheat field. This courageous but senseless act drew hundreds of Rebel riflemen, who shot down all but 71 of the 164 Union horsemen.

With night falling, Banks's corps fled in defeat northward. When advance Confederate skirmishers a mile and a half beyond the battlefield came up against a solid Federal infantry line with banks of artillery in place—Ricketts's division—Jackson called off pursuit.

Jackson had stopped an army twice the size of his own. He threw Pope's still-superior army on the strategic defensive and gained more than a week, in which time the bulk of Lee's army could come up.

Banks lost 2,381 men out of about 9,000 committed, most killed and wounded. Jackson lost 229 men killed and 1,047 wounded, nearly half in Garnett's and Taliaferro's brigades; he gained 400 prisoners, 5,300 stand of arms, and a cannon. Banks's corps had been rendered combat-ineffective, and Pope relegated it to guard duties in the campaign about to unfold.

＊   ＊   ＊

STONEWALL JACKSON SAW an opportunity to destroy John Pope's entire army of 52,000 men before Pope could be reinforced.[16] The chance opened up because of two events.

The first event occurred after Halleck ordered McClellan to withdraw the Army of the Potomac from the Peninsula. When this happened, the Union's largest and most powerful army—more than *half* the North's troops—temporarily ceased to exist as a fighting force. Since McClellan's troops had to traverse a roundabout route by water and land, the Confederate army was handed several crucial days to concentrate at Gordonsville before the Union army could unite against it.

The second event occurred when Pope placed his army in an extremely vulnerable position. Although Pope had been tempted to chase Jackson when the Confederates withdrew from Cedar Mountain to Gordonsville on August 11, Halleck restrained him. "Beware of the snare," Halleck warned. "Feigned retreats are 'Secesh' tactics." He forbade Pope to advance farther than the Rapidan River. But in stopping at the river, Pope allowed his left or eastern flank to remain woefully weak and virtually unguarded.[17] This meant that the Confederates could sweep around Pope's left, cut off his supplies and any troops coming to help, drive his now isolated force against the Rapidan or against the Blue Ridge to the west, and either destroy it or force it to surrender. In other words,

Pope's incaution had provided gratuitously the same sort of exposure that an army defeated in battle would provide the victor—the chance to swing around its flank and destroy it. Jackson drew up plans for such an operation.

General Lee, getting confirmed reports that McClellan was abandoning the peninsula, ordered James Longstreet with ten brigades to move to Gordonsville on August 13. The next day Lee decided to go himself, and he drew after him most of the remaining Confederate troops still watching McClellan. A Rebel force of about 55,000 men rapidly assembled at Gordonsville.

On August 15, Lee met with Jackson and Longstreet to work out a course of action. Jackson presented his proposal to go around Pope's eastern flank and destroy his ill-placed army. Lee accepted Jackson's plan but—without so informing Jackson—refused to adopt crucial aspects of it.[18]

Jackson wanted the Confederates to cross the Rapidan the next night, August 16, at two fords, Somerville (on present-day U.S. Route 522) and Raccoon Ford, two miles farther east. But Longstreet insisted on more time to accumulate food supplies. Jackson offered to give Longstreet enough biscuits for the march, but Lee delayed the approach to the fords until August 17 and the sweep around Pope until August 18. To Jackson, it was imperative to move quickly so that Pope would not get wind of the movement. On the morning of August 18, however, Lee postponed the assault for another day.

Lee had a number of reasons for delaying the assault. Richard H. Anderson's division, arriving from Richmond, was not yet in position; Longstreet still didn't have enough biscuits; and Jeb Stuart had not, as ordered, burned the bridge at Rappahannock Station on the Orange and Alexandria Railroad, Pope's line of supply.[19]

Stuart was not ready to move because one of his brigade commanders, Fitzhugh Lee, was a day late in getting to the Rapidan, having stopped to resupply his troops. When Stuart's adjutant, Major Norman Fitzhugh, went out hunting for the missing brigade on the night of August 17, he was captured by a Federal cavalry patrol. The Union horsemen found on Fitzhugh's person a copy of Lee's order outlining the plan of attack. The same Federal patrol also nearly captured Stuart at Verdiersville, near Raccoon Ford and a few miles east of Orange.[20] Stuart fled without his cherished plumed hat, a loss that vastly amused the Confederate army. Much more important was the seizure of his dispatch case, which told Pope that Jackson had been strongly reinforced.[21]

Robert E. Lee had put much stock in burning the Rappahannock Station bridge. To rest Fitz Lee's horses, General Lee postponed the attack once more, to the morning of August 20. Jackson protested, since the Confederates already knew the location of the Union divisions and the enemy was quiet. Indeed, the Federal army was lying placidly in camps scattered over a wide tract of country, and the cavalry was idle, seemingly unconscious of the proximity of the Confederate army. Jackson said he had enough cavalry to protect the flanks of the whole army. The purpose of the offensive, Jackson emphasized, was to turn Pope's left, move on his rear, and annihilate his army, not merely to cut his line of communications.

Lee, however, overruled Jackson.[22] Although he had accepted Jackson's plan to swing around Pope's eastern flank, he had not told Jackson that he had not accepted his concept of driving Pope's army against the Rapidan or the Blue Ridge. Lee simply wanted to force it to retreat. Lee kept this a secret from Jackson and everyone else in his army, though he did confide it to President Davis.

Therefore, the two generals went into the campaign with radically different ideas as to its purpose. Lee was fixated on breaking Pope's railroad supply line as a means of forcing him to retreat. To him, this was more important than attacking quickly. Jackson felt precisely the opposite. He wanted to strike on August 16. August 20 was impossibly late. Speed was more important than biscuits or burning a bridge. Speed was surprise, and surprise could bring victory.

Lee's delay proved fatal to Jackson's hopes of destroying Pope's army. On the morning of August 18 Pope got word of the planned attack from the order found on Major Fitzhugh and from a spy who reported that the Rebel army was assembled just south of the eastern fords. He started his army at once toward the Rappahannock River, twenty to twenty-five miles away. Progress was slow, and the Union army on the morning of

August 19 was still some distance from the Rappahannock and not in a position to resist an attack. But it was not until the afternoon of August 19 that Lee discovered the movement from an observation station atop Clark Mountain, just south of the Rapidan.

When the Army of Northern Virginia set out on August 20 in pursuit, therefore, it encountered only rear guards. That day Pope got his army on the north side of the Rappahannock, with guns mounted and ready to challenge the Confederates.

The South had lost one of the great opportunities of the war. John Pope had posted his army in an isolated position. Lee had gotten an army superior in size within striking distance. Burnside and McClellan's army were still days away. The conditions were perfect to eradicate Pope's army before having to deal with Burnside and McClellan. These forces might have been defeated in detail. Lee allowed this opportunity to pass.

One Union General grasped what could have befallen his army. George H. Gordon wrote: "It was fortunate that Jackson was not in command of the Confederates on the night of August 17; for the superior force of the enemy must have overwhelmed us, if we could not have escaped, and escape on that night was impossible."[23]

❈   ❈   ❈

ON AUGUST 20, 1862, the Army of Northern Virginia found General Pope's army arrayed along the river from Rappahannock Station (present-day Remington on U.S. Routes 15 and 29) to Kelly's Ford, four miles downstream. When the Confederates tested the fords they found them fiercely defended.

Lee realized he had only a short time to deal with Pope before the odds became impossibly high. McClellan's army was beginning to come ashore at Aquia Creek near Fredericksburg, and Lee feared that part of Pope's army would be sent to join McClellan and march directly on Richmond. He sent two divisions—D. H. Hill's and Lafayette McLaws's 17,000 men—to the North Anna River above Hanover Junction north of Richmond to stand guard. Meanwhile, he hoped to maneuver Pope's army off the Rappahannock before having to turn and deal with McClel-

lan at Fredericksburg. Accordingly, Lee ordered his army to move upstream to seek some ford around which it could sweep to displace Pope.
But the Union army kept step with the Confederate advance. In this way,
both armies, clashing along the way, moved their western flanks to Waterloo, thirteen miles upriver from Rappahannock Station.

Lee was still confident Pope would retreat if his rail connection was
broken. On August 22 he approved another cavalry strike on a railroad
bridge, this time at Catlett Station over Cedar Run, twelve miles north of
Rappahannock Station. Jeb Stuart's raid occurred during a raging thun-
derstorm that made the bridge too wet to burn. However, Pope's headquarters was there, and Stuart found Pope's dispatch book, containing
his orders.[24]

The book revealed that Pope had not detached any troops to Fredericksburg after all, but planned for McClellan to join him along the Rappahannock. Many Federal troops were already on the way. Lee had only a
few days before Pope's army would be entirely too large to confront.[25]

With the danger to Richmond gone, Lee ordered the divisions of D. H.
Hill and McLaws to join him. But he could not wait for them. On August
24, Lee told an astonished and delighted Jackson to take his whole corps,
23,000 men, get on Pope's rear, and plant it at some point on the railroad.
This move was sure to compel Pope to give up the line of the Rappahannock and rush back to reopen his supply line. After breaking the railroad,
Jackson was to hold Pope at bay until the remainder of Lee's army could
reach him. Meanwhile, Longstreet was to demonstrate along the river to
divert attention.[26] Jackson seized on Lee's idea, seeing in the general's
modest plan the chance for a much greater strategic campaign.[27]

To bring Pope to battle under conditions favorable to the South, Jackson would have to go around Pope's right or western flank across the line
of low mountains (part of the Catoctin chain but known locally as the Bull
Run Mountains) running northward into Maryland. Although Jackson
could have struck the railroad quicker by a move around the eastern
flank, Union cavalry were alert in this sector, and such a move would have
left him open to assault by Union troops coming from Fredericksburg.

Jackson called in his chief engineer, Captain J. Keith Boswell, a native of the region. He told Boswell to determine the most covered route around Pope's right. The general objective, he said, was Manassas Junction, twenty-five miles in the Federal rear and only eighteen miles from Alexandria. Jackson, then, was not planning to break the railroad at the closest accessible point; he was looking to block Pope's retreat to the capital and to challenge him to a decisive battle.

Jackson assembled his three divisions around Jeffersonton and ordered his men to cook three days' rations and to strip all vehicles from the column except ambulances and ammunition wagons. The haste was so great that some men didn't have time to prepare rations and had to rely on green corn and apples taken from fields and orchards along the way. For Jackson, there was no waiting until sufficient biscuits had been baked.

Early on the morning of August 25 Jackson's column moved out, marching first to Amissville, six miles northwest, passing around Piney Mountain, and striking north across the Rappahannock at Henson's Mill. Now behind the mountains, the men marched through Orlean and north to Salem (present-day Marshall) on the Manassas Gap Railroad. It was a hard march of twenty-six miles.[28]

The Federals discovered the movement by 8:00 A.M. and observed it for fifteen miles from a signal station on a mountain near Waterloo. Pope discussed what it meant with other generals, but no one could imagine that Lee was splitting his army. Pope concluded that Jackson was on the way to the Shenandoah Valley, and the rest of Lee's army might be following. He did not, however, send out cavalry to find out if his conclusion was correct. Instead, he ordered Franz Sigel at Waterloo and Irvin McDowell at Fauquier White Sulphur Springs, four miles downstream, to attack across the river. The attack could not be mounted until the next day, and then only McDowell was ready. If Lee was actually moving toward the Valley, it made little sense to assault across the Rappahannock.[29]

When Jackson's troops reached Salem, they collapsed in exhaustion, without bothering to unroll their blankets. Long before dawn their officers stirred them, and they moved off in the darkness. As usual, no one had any idea where they were going. But a great wave of enthusiasm passed through the ranks at dawn on August 26 when the men saw they were heading toward the rising sun. They were marching right into the heart of enemy country. Somewhere out there they were going to fall on the rear of the enemy's army. That day the Confederate column moved through White Plains (now The Plains), the Bull Run Mountains, and Thoroughfare Gap, and on toward Gainesville on the Warrenton-Alexandria turnpike (now U.S. Route 29); eight miles beyond that village lay Manassas Junction.

Pope and his major subordinates remained baffled on the Rappahannock. Pope asked Irvin McDowell to find out what had happened to Jackson's column. He also allowed McDowell to decide whether to attack alone across the river, which McDowell opted against doing. When McDowell ordered his cavalry to move after Jackson, his commander said his horses and men were broken down and he couldn't get away until the next morning, August 27.[30] By that time Jackson had more than made his presence known.

McDowell became convinced that the mass of the Confederate army was at or above Waterloo. He didn't know what this signified. But Pope—suspicious that something was up—decided to break his army away from the Rappahannock in the afternoon. This convinced Lee, watching closely, that Pope was moving back to counter Jackson.

In fact, Pope still had no idea where Jackson was. He had received reports that hostile troops had passed through Salem, White Plains, and Thoroughfare Gap, but the most he thought possible was that Lee might attempt an attack on Warrenton, six miles east of Waterloo. He never conceived the possibility that Lee would dare to divide his army or that Jackson could march fifty miles in two days and place his single corps astride Union communications.

Accordingly, Pope ordered three of his corps—McDowell's, Franz Sigel's, and Jesse L. Reno's of McClellan's army—to concentrate the next day, August 27, at Warrenton. He left Nathaniel P. Banks's still-shaky small corps at Fayetteville (now Opal), seven miles south of Warrenton, and directed Fitz John Porter's corps of 10,000 men, just arriving, to move toward Warrenton Junction (now Calverton) on the railroad, nine miles southeast of Warrenton. Samuel P. Heintzelman's corps was already located there. Pope now had 75,000 men.[31]

When Stonewall Jackson's corps marched into Gainesville around

4:00 P.M. on August 26, Jeb Stuart and his cavalry arrived as well. They had left Waterloo at 2:00 A.M. and traveled by back ways through the Bull Run Mountains. Despite Stuart's and Jackson's movements, and despite the high columns of dust they raised, the Federal army remained unaware of the presence of a major force on its rear.

With Stuart now screening the march, Jackson directed the column to move on Bristoe Station, four miles south of Manassas Junction. The head of the column reached Bristoe at sunset and quickly dispersed a small Federal guard. The Confederates failed to stop the first locomotive coming north from Warrenton Junction with a load of empty cars. It rushed on to Manassas Junction to warn the authorities. But the Rebels cut the telegraph wires and tore up track, which wrecked the second and third trains coming north. The fourth train stopped and backed up to warn Pope. The Confederates now moved to Broad Run and broke the bridge that crossed there. This had been Jackson's objective, to cut off the Federals' supply line and force them to retreat.[32]

People at Bristoe reported that Pope's main supply base was at Manassas Junction—acres upon acres of goods of all kinds lying alongside the tracks and in warehouses. Although the Confederate army was exhausted, Colonel Isaac R. Trimble volunteered to lead the 21st North Carolina and 21st Georgia, 500 men, up the tracks to secure the junction. Meanwhile, Stuart sent the 4th Virginia Cavalry around to the rear of Manassas. The Rebels overcame feeble Union resistance in minutes, capturing more than 300 prisoners.

Pope got word of the action at Bristoe soon after it happened. But he thought it was a hit-and-run strike by a small group of cavalry and ordered Heintzelman to send a regiment on a train from Warrenton Junction to repair the wires and protect the railroad. The regiment sent to Bristoe returned with the news that the enemy was not a small body of horsemen but a strong infantry force.

Now thoroughly alarmed, Pope at 7:00 A.M. on August 27 sent Joseph Hooker's division toward Bristoe from Warrenton Junction to do what the regiment had failed to do. At 8:30 A.M. he directed most of his army back, not to Warrenton but to Gainesville, twelve miles northeast. Jackson's thrust had achieved its original purpose: Pope abandoned the Rappahannock line.

Most of the remainder of Lee's army also was on the way north. Convinced on the afternoon before (August 26) that Pope was withdrawing, Lee had pulled all but Richard H. Anderson's 6,000 men off the river and set out on Jackson's roundabout route. By the morning of August 27 Lee and Longstreet were well on their way toward a junction with Jackson.[33]

When the bulk of Jackson's corps moved up on Manassas Junction on the morning of August 27, they found a fairyland of stores of all kinds. Every item of clothing and any food, from luxuries such as champagne, Rhine wine, coffee, and canned sardines to everyday rations, were there for the taking for the famished and ill-supplied Rebels. Jackson left Ewell's division at Broad Run and Bristoe Station to block any aggressive move up the railroad from Warrenton Station.

At Manassas, the Stonewall Brigade quickly dispersed the 12th Pennsylvania Cavalry, sent down from Alexandria. Soon thereafter, the inexperienced 1st New Jersey Brigade, sent by Halleck, detrained just north of the railroad bridge over Bull Run. Its commander, Brigadier General George W. Taylor, had been told only to protect the bridge. But he decided to press on to Manassas Junction against what he thought was a small Rebel force.

The brigade ran headlong into a large part of A. P. Hill's division. Realizing that it was about to be destroyed, Jackson himself rode forward and called on the Federals to surrender. He received a bullet past his ear in reply. The Rebel artillery now opened on the Union brigade and broke it apart. More than 300 Federals surrendered, while 135 were killed and wounded. Taylor himself died. The Confederates destroyed the bridge over Bull Run and the locomotive and cars that had brought the brigade, and Fitz Lee's cavalrymen pursued the fugitives all the way back to Burke Station, twelve miles from Alexandria.

Jackson knew he could not remain for any length of time at Bristoe and Manassas. All through the afternoon Stuart's patrols sent Jackson report after report on enemy movements. A large force was marching on Bristoe from Warrenton Junction; heavy columns were heading down the turnpike from Warrenton toward Gainesville. Meanwhile, a courier disguised as a countryman arrived from Lee, informing him that Lee and Longstreet were following Jackson's own track.[34]

On the evening of August 27, Pope arrived at Hooker's division, which was now facing Ewell just south of Bristoe Station. By marching his army along the Warrenton-Gainesville road, he had arrived in the central position between two weaker wings of the Confederate army. Even if he merely concentrated his army at Gainesville, he could keep Lee's two wings apart. Jackson would be isolated east of Gainesville, Longstreet west. (Longstreet's only feasible entrance eastward was Thoroughfare Gap or other gaps nearby, all of which Pope could block.)

But Pope did not see this opportunity. Discovering that Stonewall Jackson's whole corps was occupying Manassas, he concluded that Jackson's entire purpose had been a raid. After all, Lee had already attempted to burn the railway bridge in a raid on Catlett Station on August 22. But Pope then came to the remarkable conclusion that Jackson was trying to escape southward around Pope's *eastern* flank and that Hooker had him blocked. Pope completely ignored the fact that the rest of Lee's army was approaching from the west toward Thoroughfare Gap and that Jackson's logical sally port was *west,* toward Lee and away from Hooker.

Nevertheless, Pope decided that no other course was open to Jackson but to entrench himself at Manassas Junction and await Lee's arrival. He sent out orders for all his forces to march at once on Manassas. "We shall bag the whole crowd," Pope said, if his forces moved up promptly.[35] He abandoned his central position between Lee's two wings. Meantime, Ewell, in a series of adroit moves, withdrew north of Broad Run and marched up to Manassas Junction.

So on the morning of August 28, 1862, the Federal army began frantic marches on Manassas Junction with the expectation that Jackson was cornered there. It was noon before Hooker's advance guard arrived at Manassas. The huge cache of Union stores had been consumed by fire. Smoking ruins greeted the Union soldiers. Jackson and his corps had seemingly vanished.

Second Manassas

A NUMBER OF THINGS had gone badly wrong very quickly for John Pope. The Confederates had knocked him off the Rapidan and Rappahannock rivers with nothing but guile. His men were running out of food because of the destruction of his supply depot. The force that he had confidently expected to evict from Charlottesville only three weeks before was within twenty miles of the national capital—the very place President Abraham Lincoln and Secretary of War Edwin Stanton had told him was his primary job to protect.

General Pope had abandoned a superb strategic position between the two wings of the Rebel army. He had turned his back on the largest part of the Confederate army under Robert E. Lee. He had failed to see that the Rebel force that had struck Manassas Junction was entirely too large for a hit-and-run raiding party and must have some other purpose. And he had directed all of his forces on Manassas in a move that required the intended victim there, Stonewall Jackson, to sit forlornly and await his destruction. It is difficult to imagine how General Pope could have gotten any more things wrong—but he did.

When the advance guard of the Union army arrived at the devastated Manassas station at midday on August 28, Pope was completely bewildered. For four hours he had no idea what had happened to Jackson or what to do. Then at 4:00 P.M. news arrived that the rail line at Burke, just twelve miles from Alexandria, had been cut and that Confederates were in force between Burke and Centreville. Thinking now his quarry was fleeing in disorder, Pope directed his whole army on Centreville, twelve miles north of Manassas and just above the Bull Run battlefield of the year before.

But when leading elements of the army reached Centreville, there were no Confederates there. The only evidence of the enemy was a few Rebel horsemen who retreated leisurely before the advance. The Union soldiers, who had marched to no avail all day, dropped down wearily. Then, seven miles away across Bull Run near Groveton on the Warrenton Turnpike, the dull boom of cannon erupted, swelling to a continuous roar. The Union soldiers at Centreville could clearly see the rolling smoke of battle above the woods along the turnpike. Still farther westward, where the cleft of the blue hills marked Thoroughfare Gap in the Bull Run Mountains, the flash of distant guns lit the sky.

Jackson, the soldiers agreed, had been run to ground at last. Irvin Mc-Dowell's forces, moving on Centreville from Gainesville, had cornered him. But this facile conclusion had flaws. Jackson now was within the reach of Lee, for those guns in Thoroughfare Gap most assuredly were marking the advance of the main Confederate army toward Gainesville. And there was something disquieting about how Jackson had slipped out of the net at Manassas. How had 23,000 men, with batteries and wagons, passed without having been seen through the cordon that General Pope had so confidently thrown around the station?

The answer, of course, was that Jackson had marched well before Pope had set his snare. On the night of August 27, while Pope, Hooker, and their troops were watching the red glare of the fires rising from Manassas, Jackson's corps was moving north on three roads. Before dawn A. P. Hill was near Centreville, Richard Ewell had crossed Bull Run at Blackburn's Ford, and William B. Taliaferro, going straight up the Sudley Springs–Manassas road with the trains, was north of Henry House Hill, scene of battle the year before. Taliaferro continued on until his division was hidden in the woods above Groveton, some eight miles northwest of Manassas.

Jackson didn't know that Pope was trying to "bag" him at Manassas on August 28. He knew, of course, that he could not remain at the junction. He had to reach a position from which he could retreat if disaster

struck, and where he could reunite quickly with Lee. The site at Grove-
ton was ideal. It was close to Thoroughfare Gap, and in case of defeat
Jackson could move north through nearby Aldie Gap in the Bull Run
Mountains.

There was another reason Jackson selected Groveton. His purpose in
descending on Manassas Junction was not merely to break Pope's supply
line but to place his own corps in such a seemingly exposed position that
Pope would be incited to attack him. He also hoped to prevent Pope from
trying to flee back to Washington along the Warrenton Pike, as the Union
general had already run twice. That's why he sent A. P. Hill and Ewell
north of Bull Run—to block Pope's passage to the defenses of Washington. Jackson wanted Pope to stay in northern Virginia because he wanted
to destroy his army.

Jackson knew the landscape around Groveton. His brigade had occupied the neighborhood after First Manassas. At the time Jackson had
found a tree-covered ridgeline above Groveton, three miles east of Gainesville and about a mile north of the Warrenton Pike. The primary attraction
of Groveton Heights was its suitability for defense, having an unfinished
railroad line running just under it, with deep cuts that could serve partially as defensive emplacements.

Groveton Heights also was an admirable site to test Jackson's new
theory about defensive battles. It had an open western flank at Gainesville, around which an enveloping force might strike. It commanded the
turnpike, the best retreat route for the Northern army back to Washington. And the Stone Bridge over which the Federals had fled at First Manassas the year before was the only adequate crossing of Bull Run in the
vicinity.

If the Union army could be enticed into attacking Jackson along
Groveton Heights, it could be defeated, and the Confederates might then
sweep around the western flank, pin the Federal army against Bull Run,
and destroy it.

*   *   *

ON THE MORNING of August 28 Confederate cavalry captured a Union
courier who was carrying orders directing Franz Sigel at Gainesville to

march to Manassas Junction. This was part of Pope's assemblage of forces to bag Jackson. But Jackson, not believing that Pope could be so foolish as to think he was still at Manassas, concluded that his position at Groveton had been detected and that Pope was fleeing to Washington by way of Manassas. He sent orders to A. P. Hill to block the fords of Bull Run to keep Pope from getting away.

But Hill had captured another Union courier ordering McDowell to prepare for battle on the Manassas plains near the junction. This proved that Pope was trying to destroy Jackson, not run. Hill ignored Jackson's order and marched to Groveton to join Taliaferro and Ewell, who had already gone ahead.

This last message eased Jackson's anxiety, leading him to believe that Pope was ripe to be drawn into battle. During the afternoon, however, he got reports from Jeb Stuart's horsemen that Pope was moving his whole army toward Centreville. Once more Jackson feared that Pope had decided to run.

One incident reinforced this concern. Confederate Colonel Bradley T. Johnson, commanding a brigade closest to Gainesville, discovered a column that he thought was advancing on him. Actually, it was John F. Reynolds's Pennsylvania division, which was still trying to get to Manassas. Johnson opened fire, and Reynolds brought his superior artillery into action. Johnson, overawed, promptly withdrew. Reynolds concluded that the Rebel force was only a small reconnoitering party and continued on toward Manassas. Jackson, however, feared that Reynolds had been trying to clear the Warrenton Pike for Federal forces to march on Centreville.

Jackson resolved to stop this movement. He also believed that if he could bring a part of Pope's force to battle, Pope might be provoked to attack. Accordingly, Jackson marched Taliaferro's and Ewell's divisions to a point slightly north of the turnpike at Groveton and close to Gainesville.[1]

Around 5:00 P.M. Rufus King's division of 10,000 men in McDowell's corps started down the Warrenton Pike on the way to Centreville. Near sunset it came opposite Jackson's position.

To Jackson this was confirmation of his fears, and he wasted no time.

He ordered the infantry to attack, and brought forward artillery, though the ground was difficult and only two batteries got into action. King's division was caught by surprise but reacted quickly. The battle of Groveton, as it was called, turned into a bloody, stand-up fight fought largely by brigadiers on both sides who fired salvo after salvo into each other's ranks. Only at darkness, around 9:00 P.M., did King's division at last fall back. The Federals lost about 1,500 men, the Confederates more than a thousand. Both Taliaferro and Ewell were wounded, Ewell losing a leg. Command of Taliaferro's division passed to William E. Starke, Ewell's division to Alexander R. Lawton.

The battle at Groveton was precisely the kind of engagement Jackson was trying to avoid. But it served his purposes, for it showed Pope where he was and challenged the Union general to battle.

The sounds of the collision caused Franz Sigel to turn his corps and John F. Reynolds his division toward Groveton. Both got within a mile of King and bivouacked for the night.

❊   ❊   ❊

WHILE THE GROVETON fight was going on, another engagement was being fought at Thoroughfare Gap. Pope, in his frenzy to bag Jackson, had completely forgotten about Lee and James Longstreet coming through the gap. But McDowell told one of his division commanders, James B. Ricketts, to watch for Longstreet. Ricketts arrived in Thoroughfare Gap just as Longstreet's corps came up in midafternoon on August 28.

Longstreet pushed David R. Jones's division forward on the main road and sent a strong flanking force of two brigades through Hopewell Pass, three miles north. Ricketts deployed one brigade forward into Thoroughfare, but Jones's men seized the heights on either side and drove Ricketts back to the eastern foot of the pass. Hearing reports that a Rebel force was approaching his flank, Ricketts withdrew, backing up toward Gainesville.

Thus on the night of August 28, Longstreet's corps was bivouacked on both sides of Thoroughfare Gap, and Jackson's corps rested at Groveton. The Federal army was spread over the map. King was at Groveton facing Jackson, Reynolds and Sigel were within supporting distance of King on the east, Ricketts was nearing Gainesville, Philip Kearny's division and Jesse Reno's corps were at Centreville, Joseph Hooker's division was just north of Manassas, Fitz John Porter's corps was at Bristoe Station, and Nathaniel Banks's corps was with the army trains at Kettle Run, about two miles south of Bristoe.

※   ※   ※

GENERAL POPE DREW all the wrong conclusions from the clash at Groveton. He decided that Jackson was trying to flee through Thoroughfare Gap to rejoin Lee. He believed that Ricketts and King were standing fast *west* of Jackson around Gainesville, and that McDowell, in command, could fend off Longstreet while Pope brought up his other forces and destroyed the isolated Jackson. At 9:20 P.M. on August 28 he ordered McDowell to hold his ground at Gainesville "at all hazards and prevent the retreat of Jackson to the west." To his other corps he wrote: "McDowell has intercepted the retreat of the enemy, Sigel is immediately in his front, and I see no possibility of his escape."

But King and Ricketts were about to remove themselves entirely from the scene, and McDowell was not in command. He had gone to Manassas to see Pope, had missed him since Pope had ridden to Centreville, and couldn't find his way through the woods back to Gainesville. So Pope's orders never got to him. As the night wore on, King grew increasingly nervous. Around 11:00 P.M. he learned that Ricketts was falling back from Thoroughfare. This was the last straw. King decided he was exposed and set his division marching for Manassas—at 1:00 A.M. Ricketts, feeling isolated at Gainesville, also retreated down the road to Bristoe.[2]

Having no idea that King and Ricketts were abandoning their splendid positions between the two wings of the Confederate army, Pope directed his other forces to converge on the supposedly fleeing Jackson at Groveton early on the morning of August 29. Since his army had spread out in all directions on August 28, it would have been better to have concentrated forces at Centreville and waited for reinforcements before attacking. But Pope believed he had sufficient strength to destroy Jackson.[3]

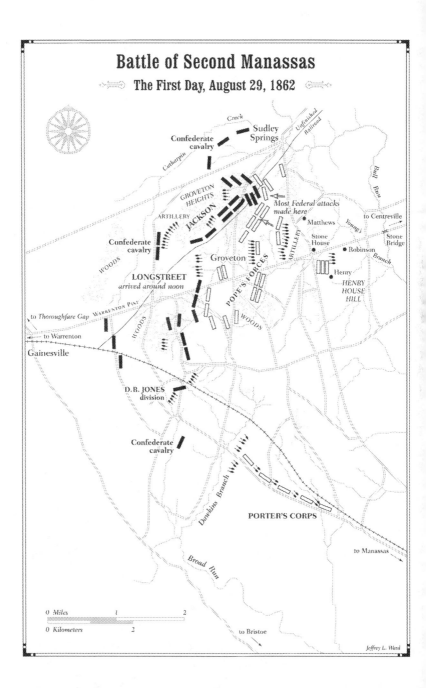

Battle of Second Manassas
The First Day, August 29, 1862

Creek

Sudley Springs

Confederate cavalry

Unfinished Railroad

Catharpin

GROVETON HEIGHTS

ARTILLERY   JACKSON

Most Federal attacks made here

Matthews

• Stone House

to Centreville

Ball Run

Young's

Stone Bridge

Confederate cavalry

Groveton

WOODS

LONGSTREET
arrived around noon

POPE'S FORCES

ARTILLERY

• Robinson

Branch

• Henry

HENRY HOUSE HILL

to Thoroughfare Gap   WARRENTON PIKE

to Warrenton

WOODS

WOODS

Gainesville

D.R. JONES
division

Confederate cavalry

Dunkins Branch

PORTER'S CORPS

to Manassas

Broad Run

0 Miles   1   2

0 Kilometers   2

to Bristoe

Jeffrey L. Ward

He felt it was important to attack Jackson at once, before the Confederate leader could flee. But there undoubtedly were other reasons. As the English Colonel G. F. R. Henderson writes, "Ambition, anxiety to retrieve his reputation, already blemished by his enforced retreat, the thought that he might be superseded by McClellan, whose operations in the Peninsula he had contemptuously criticized, all urged him forward. An unsuccessful general who feels instinctively that his command is slipping from him, and who sees in victory the only hope of retaining it, seldom listens to the voice of prudence."[4]

On the morning of August 29 McDowell found his way out of the woods, discovered what had happened to Ricketts and King, and informed Pope that his left wing had evaporated. Pope thereupon ordered McDowell with his two divisions and Porter with his corps to march on Gainesville from Manassas. They were, however, more than eight miles away and could not beat Longstreet to it.[5]

Pope still might have pulled off a victory if he had insisted that Porter and McDowell, who had about 24,000 men, attack Longstreet at Gainesville and hold him while he evicted Jackson by turning his left, or eastern, flank at Sudley Springs. Even without Porter and McDowell, Pope had more than twice as many men as Jackson. But Pope could conceive no other tactical idea except to crash headlong against Jackson's front. And he gave McDowell and Porter an excuse to back off Longstreet by authorizing them to depart from his orders "if any considerable advantages are to be gained."[6]

Around daylight Pope sent a directive to Sigel to attack Jackson at once, and for Reynolds to support him.

During the night Jackson had pulled his corps up on Groveton Heights. He was occupying a short line two miles long on elevated land suitable for artillery, running along cuts of the unfinished railway from Sudley Springs on the east to a point above the turnpike a half mile west of Groveton. It was shielded on both flanks by Jeb Stuart's cavalry. The position was one and a quarter miles from the turnpike on the east and a quarter of a mile from it on the west. Thus the line ran northeast-southwest, at an angle to the approximately east-west turnpike, along which Union forces were lining up to attack. A. P. Hill was on the left, or northeast, Lawton in the middle, and Starke on the right. The troops were largely positioned in the woods. In front of them the ground was open, except for woods, four hundred to six hundred yards wide, mostly along Hill's front, but also extending into Lawton's.

The shortness of the line gave Jackson five rifles to the yard and allowed him to concentrate five artillery batteries on the heights. In addition, he gave each local commander authority to bring up reinforcements on his own authority to any point he saw threatened. Each brigadier commanded his own reserves in a second line directly behind the first. Jackson also gave to each division commander half of his entire force to form a third line to be thrown forward whenever danger appeared. With these dispositions, Jackson expected to prove that a defended line could not be broken by a frontal attack.

The Federal attacks struck almost entirely Hill's division on the left. Although it was farthest from the turnpike, Hill's position could be approached through woods, giving a degree of protection. On the Federal left, or west, the Union troops largely contented themselves with skirmishing and artillery duels, and accomplished nothing.

On the extreme east near Sudley Springs some skirmishers from Carl Schurz's division of Sigel's corps got around Hill's flank, but Stuart's cavalry drove them away. A greater threat to the Confederates appeared in front of Maxcy Gregg's South Carolina brigade.[7] There Schurz's main body advanced through the woods. Gregg's men rushed into the timber, flanked the "Dutchmen," as the Rebels called Schurz's Germans, and in a violent clash split two Federal brigades apart and broke the division's center. Schurz's men rallied, but again Gregg's brigade drove the Federals out of the woods in disorder.

Schurz's division tried again, reinforced by other foreign soldiers from Adolph von Steinwehr's division. Here a violent battle raged in the railway cut, the "Dutchmen" making a small penetration but falling back in disorder when two Rebel regiments struck their flanks. Schurz's division pulled away about noon, exhausted and demoralized.

While these bitter fights were going on, Lee and Longstreet marched down from Thoroughfare Gap, the main body arriving at Gainesville about 11:30 A.M. Longstreet turned east along the turnpike and drew up on either side of it, facing approximately east and at a near right angle to Jackson's line.

At about the same time Fitz John Porter arrived at Dawkins Branch, about four miles south of Gainesville, with his two divisions and King's, about 17,000 men. McDowell joined him there at noon. They heard from Federal cavalrymen that Longstreet's van had passed through Gainesville, and they could see dust clouds, indicating that the whole corps was marching to the aid of Jackson.

Here was a chance to strike a telling blow before Longstreet could get into position, using the 17,000 men in place and bringing up Ricketts's 7,000 men from Bristoe and Banks's 10,000 from Kettle Run. But neither Porter nor McDowell suggested an attack. They thought only of making contact with the rest of the Union army on the turnpike. They concluded that a direct move across country from Dawkins Branch was not feasible because the terrain was broken. Instead they decided that Porter should remain where he was and that McDowell would take King's division over to the Sudley Springs Road and march along it to the Warrenton Pike, with Ricketts's division to follow. In other words, McDowell would march two divisions from the *flank* of the Confederate army in order to attack that army directly *in front*. And two entire divisions would remain idle at Dawkins Branch.

Along the turnpike Pope prepared to strike a massive blow against Jackson's position as soon as he heard of McDowell and Porter's advance on Gainesville. Pope had received no word of the arrival of Longstreet, McDowell and Porter having failed to inform him. Nor had they told him that McDowell was moving away from Dawkins Branch and that Porter had no plans to do anything.

To support his advance, Pope planned two simultaneous attacks on the turnpike, one by Joseph Hooker, the other by Philip Kearny's and Isaac I. Stevens's divisions. Hooker selected only a single brigade, Cuvier Grover's Massachusetts–New York–New Hampshire–Pennsylvania outfit, to conduct the attack. Grover ordered his soldiers to load their rifles, fix bayonets, and move slowly until they felt the enemy's fire, then to deliver a volley, charge, and carry the position by main force, relying on the bayonet.

Grover's brigade went up against Edward L. Thomas's and Maxcy Gregg's now exhausted forces. The Union soldiers marched up to the railway cut, delivered the volley, and rushed forward, driving the stunned Rebels beyond the railroad line. But Confederate artillery staggered the Federals, and Rebel reserves hurried down and broke the valiant force. More than a quarter of Grover's 2,000 men were killed, wounded, or captured.

Kearny's and Stevens's attacks did not go in at the same time as Grover's, as ordered. The two Union commanders waited until Grover had failed, then lined up at 5:00 P.M. and prepared to assault Gregg's brigade, which had already lost a third of its 1,500 men and was sagging with fatigue. When the Federal storm burst against Gregg, the South Carolinians fell back to a last-ditch stand on the hill behind the railway cut.

But Hill had in reserve Jubal A. Early's Virginia brigade, plus the 13th Georgia and the 8th Louisiana, and sent them in. Early struck the Federal center with a single, shattering blow. The line crumpled; the Rebels rushed through, and the two Union divisions broke and ran back over the railroad cut and beyond, pursued for some distance by Early's triumphant soldiers.

At 4:30 P.M., just as the final assault against Jackson's line was about to go in, Pope had sent a peremptory order to Porter to attack the Confederate flank and rear. Porter didn't get the order until 6:30 P.M., and decided correctly not to launch it, since Longstreet had more than 32,000 men and Porter only 10,000.[8]

Longstreet, though, remained idle all day. Lee asked Longstreet three times whether he was ready to attack on Pope's left flank, and three times Longstreet said no. Longstreet reasoned that Porter at Dawkins Branch might attack, while in reality Porter's corps remained in column formation on the road and did not pose a threat. Though Lee ascertained by his own observation that the Federal force at Dawkins Branch numbered only about 10,000 men, he accepted Longstreet's decision, indicating that he was still unwilling to fight a general engagement with Pope.[9] As a result, more than half of the Confederate army remained idle on August 29, while Jackson stood alone, absorbing assault after assault.[10]

Jackson was confident that his troops could stop the attacks and deeply demoralize the Federal soldiers. Therefore, the situation was ideal for Longstreet to swing on the left flank of this disabled army, seize Henry House Hill just south of the Stone Bridge, and cut off Federal retreat over the Stone Bridge on the Warrenton Pike. In this way, the entire Union host would be forced to surrender.

Yet it was not to be. Longstreet refused to attack, and Lee didn't press him.

English General Sir Frederick Maurice is highly critical of Lee and Longstreet for this failure. "While Pope was bringing his corps up piecemeal to attack Jackson, Longstreet stood stretching well beyond the Federal left, ready to strike a deadly blow," Maurice writes. "On this day Jackson had to play the part of [the Duke of] Wellington at Waterloo, and stand the pounding until Longstreet, in the role of [Gebhard Leberecht von] Blücher, gave the *coup de grâce*. But the *coup* was not given. . . . It may be the experience of Malvern Hill weighed heavily on him, but it is clear that he was obsessed by one idea. He believed the recipe for victory to be to maneuver an army into a position such that the enemy would be compelled to attack at a disadvantage, and there await the blow."[11] Jackson believed, in contrast, that a defense-only policy would gain no decision. For him the correct course of action—after defeating an enemy attack—was to attack around the flank of the demoralized enemy. For Jackson the moment to launch a flank attack was *while* Pope was fruitlessly assaulting him on August 29. He had been successfully defending all day, and Pope's left flank was bare.

The first day of Second Manassas ended. Pope had assembled 60,000 of his 75,000 men and had to deal only with Jackson's 22,000 men. Yet Jackson's new-model defensive system shattered all Pope's attacks. Pope himself contributed to the failure by launching one disconnected assault after another. Jackson was able to concentrate more men than Pope at every critical point. He had used the railroad embankment as a defensive line, which showed that he was moving toward field fortifications. But he did not follow this idea to its logical conclusion and order his men to build and use entrenchments and breastworks. In the partially stand-up fights and counterassaults that his troops accordingly carried out, Jackson sustained heavy casualties, though only a quarter of the 6,000 suffered by Pope. And Jackson's force was never in danger.

❦   ❦   ❦

THE NEXT DAY, Saturday, August 30, 1862, General Lee entertained no plans to attack Pope and soon concluded that the Federals would retreat. But John Pope had decided to attack again. He not only believed he had won the first day's battle, but he also thought Jackson was retreating.

Pope fell for one of the oldest deceptions in warfare: during the night, Jackson had pulled his men off the unfinished railroad line back into the woods on the crest of Groveton Heights to mystify the Federals and also to give his troops a needed rest.[12] Without checking further, Pope concluded that Jackson's absence from the front lines meant he was fleeing—and that the Union had therefore won the battle the day before. Also, he continued to ignore Longstreet's corps sitting on his flank. He ordered Porter's corps to move from Dawkins Branch on his left to the Warrenton Pike, thus removing the single excuse Longstreet had given for failing to attack the day before.

Napoléon Bonaparte said that the general ignorant of his enemy's strength and dispositions is ignorant of his trade. Pope was proof of this dictum.

At 5:00 A.M. on August 30, Pope wired Halleck: "We fought a terrific battle here yesterday with the combined forces of the enemy, which lasted with continuous fury from daylight to dark, by which time the enemy was driven from the field, which we now occupy. Our troops are too much exhausted yet to push matters, but I shall do so in the course of the morning, as soon as F. J. Porter comes up from Manassas. The enemy is still in our front, but badly used up. We have lost not less than 8,000 men killed and wounded, but, from the appearance of the field, the enemy lost at least two to one. . . . The news just reaches me from the front that the enemy is retiring toward the mountains; I go forward at once to see."[13]

Pope resolved to attack the "retreating" Jackson by sending Ricketts's division around Sudley Springs on Jackson's extreme northeastern flank,

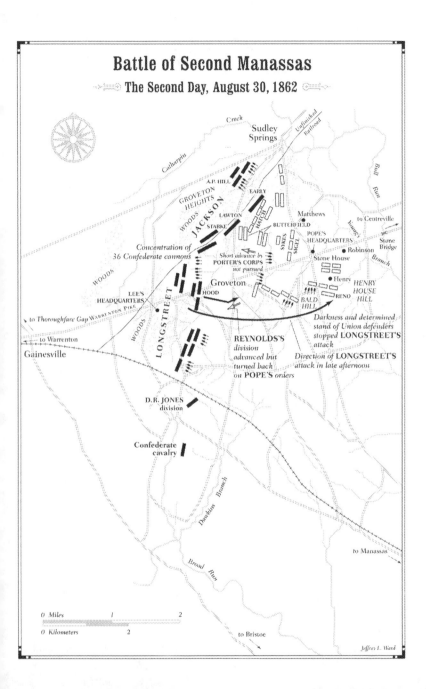

Battle of Second Manassas
The Second Day, August 30, 1862

Catharpin Creek

Sudley Springs

Unfinished Railroad

Bull Run

to Centreville

A.P. HILL

GROVETON HEIGHTS

EARLY

WOODS

LAWTON

JACKSON

STABLE

Matthews

BUTTERFIELD

POPE'S HEADQUARTERS

Concentration of 36 Confederate cannons

Short advance by PORTER'S CORPS not pursued

HATCH

SYKES

SIGEL

Robinson

Stone Bridge

Stone House

Vinards

WOODS

Groveton

Henry

HENRY HOUSE HILL

LEE'S HEADQUARTERS

HOOD

LONGSTREET

BALD HILL

RENO

to Thoroughfare Gap WARRENTON PIKE

WOODS

Darkness and determined stand of Union defenders stopped LONGSTREET'S attack

to Warrenton

Gainesville

REYNOLDS'S division advanced but turned back on POPE'S orders

Direction of LONGSTREET'S attack in late afternoon

D.R. JONES division

Confederate cavalry

Dawkins Branch

to Manassas

0 Miles   1   2

0 Kilometers   2

Broad Run

to Bristoe

Jeffrey L. Ward

while ordering Porter to push his corps westward along the Warrenton Pike toward Gainesville and around Jackson's right flank.

The sweep around Jackson's left flank collapsed the moment Ricketts discovered that Jackson was still in force along the unfinished railway. Porter did not march far along the turnpike because he concluded that no pursuit was possible until Jackson had been defeated along his front. Porter proposed instead to attack head-on into the center and right or western parts of Jackson's line along the railway "to develop the strength of the enemy." Then, Porter said, he could wheel around Jackson's right flank.[14] Like Pope, Porter ignored Longstreet's corps, sitting directly on Jackson's right—the very same force that he had decided *not* to attack from Dawkins Branch the day before. Porter also revealed that he did not know the purpose of a turning movement, which is to *avoid* the enemy's main force, not crash headlong into it. But McDowell, assigned tactical command that day, approved Porter's plan, and Pope backed him.

Porter's plan of attack overlooked the fact that Jackson's and Longstreet's corps together formed a flat crescent, meaning that any attack into Jackson's center or right would send the Union troops into a cul-de-sac, with both entry and exit blanketed by fire from two directions. The danger was greatly heightened because two Confederate artillery battalions—thirty-six cannons—were mounted on an eminence just north of the turnpike at the juncture of Jackson's and Longstreet's lines. These cannons faced generally east over a wide area of open fields in front of the center and right wing of Jackson's line—the very spots Porter had picked to assault. Their only obstacle was a belt of trees about 1,300 yards away in front of Jackson's left wing (the woods that had sheltered the Federal attacks on August 29). As Confederate artilleryman Porter Alexander wrote, "A battery established where it can enfilade others need not trouble itself about aim. It has only to fire in the right direction and the shot finds something to hurt wherever it falls. No troops, infantry or artillery, can long submit to enfilade fire."[15]

Pope's inability to read the plain facts had given General Lee another day of opportunity. With Porter gone, Lee faced no danger whatsoever on Pope's exposed left flank. Despite Longstreet's failure to attack the day before, prospects for destroying Pope's army remained as bright as ever. But Lee decided not to attack, and to hold to his original aim of avoiding a general engagement and relying on maneuver to force Pope from northern Virginia—this *after* Jackson had spent an entire day in an intense general engagement with Pope's army.

Porter lined up 20,000 Union troops behind the belt of trees that had sheltered the Federal attacks the day before. On the right was John P. Hatch's division, and on the left were two brigades of George W. Morell's division, commanded by Daniel Butterfield, because Morell had gone by mistake with one of his brigades to Centreville. To Butterfield's left rear was George Sykes's division, while Franz Sigel's corps was kept in reserve. The two assault divisions gave Porter only a 30 percent superiority of force against Jackson's troops. Half of Pope's army remained behind in column formation.

Around 1:00 P.M., Pope at last made a gesture toward protecting his left flank. He sent Reynolds's division westward along the south side of Warrenton Pike to block any Confederate move from that direction. Just southwest of Groveton, Reynolds ran into a hornet's nest of Rebels and sent back pleas for help. Instead, McDowell ordered two brigades of Ricketts's division to Henry House Hill, a mile and a half east of Groveton, while Pope sent Nathaniel C. McLean's brigade of Schenck's division to Bald Hill, a mile east of Groveton. Neither force provided Reynolds any help.

Then, seeing Jackson's first line bound out of the woods and reoccupy the railroad line just before Porter launched his attack, Pope directed Reynolds to *abandon* his position on the left flank and to move north of the pike to the rear of Porter.[16] Pope remained to the last oblivious to the danger on his flank.

At 3:00 P.M. Porter's attack moved forward. Since Jackson's line extended northeast-southwest and since Porter had turned his assault to move directly west, it struck Jackson's line at an acute angle, Hatch being much closer to Jackson's line than Butterfield was.

Hatch's lines advanced through the woods that extended to the railroad, largely protected from Confederate artillery fire from the elevation on the flank. Hatch's division struck mainly Marcellus Douglass's and Jubal Early's brigades on the left sector of Ewell's division (now under Alexander R. Lawton). The fighting was severe, but the Confederates finally repulsed the Federals when Dorsey Pender's brigade charged into the melee to restore the Rebel line.

In front of Butterfield the woods were only about a quarter of a mile deep, and beyond them stretched about a half mile of open fields. The first of Butterfield's three lines of battle emerged from the woods, but Confederate artillery could not fire quickly enough to halt Butterfield's first line as it rapidly moved across the fields to the railroad line. There Ewell's men and Taliaferro's division (under William E. Starke) were waiting. The battle at once became wild and intense. To the right, the Federals pushed back the first line of the Stonewall Brigade. Colonel W. S. H. Baylor, commanding the Rebel brigade, led the second line forward and drove Butterfield's men back, though Baylor died at the head of his men.

The battle raged in unbelievable fury for twenty to thirty minutes. There was no thought of retreat by either side. The Federals made assault after assault against the railway line, but the Rebels stopped every one. Some of the Rebels, out of ammunition, ran out and grabbed the cartridges off dead bodies or snatched up rocks and threw them at the enemy. One of the Confederate commanders under assault, Colonel Bradley Johnson, reported: "I saw a Federal flag hold its position for half an hour within ten yards of a flag of one of our regiments in the cut and go down six or eight times, and after the fight one hundred dead were lying twenty yards from the cut, some of them within two feet of it."[17]

The Confederate artillery, unable to fire on the railroad cut for fear of hitting friends, concentrated on the woods from which the succeeding Federal lines had to come and on the fields over which they had to pass to reach Jackson's line. The remainder of Butterfield's division, penned in the open between Longstreet's line and the grove of woods, was riddled with rifle fire in front and a hail of canister and shells from the thirty-six guns on the hill. The Union soldiers could not endure this inferno and fell back across the fields in disorder. Seeing that their supports behind had fled, the first line of Union soldiers at the railroad line rushed to the rear as well.

As soon as General Lee witnessed the collapse of the Federal attack, he recognized the opportunity. Casting aside his resolve to stand on the defensive, he ordered a general advance. Longstreet had anticipated the order and already had summoned his troops to charge. As fast as the word could be delivered, unit by unit the Army of Northern Virginia sprang forward in a great sweeping advance all along the four-mile front, crimson banners to the fore and the men screaming the Rebel yell.

Since Pope had ordered Reynolds away, the Federal army had no shield behind which it might have rallied. Only a few small forces were in place to contest Longstreet's advance. The Federals' greatest asset now was time. The Confederates had only a few hours of daylight to seize Henry House Hill, which dominated the Stone Bridge, the only feasible route over which the Federal army could retreat.

John Bell Hood's division led the Rebel attack, bowling over Reynolds's third brigade, which was still south of the turnpike. But at Bald Hill he was stopped by Nathaniel McLean's brigade and the two brigades of James Ricketts's division that had rushed over from Henry House Hill. By this time the Confederates were drooping with fatigue, having pushed hard and fast for three miles. Nathan George "Shanks" Evans's South Carolina brigade was unable to carry the hill on the first assault. This delay foretold the doom of the Confederates' effort to destroy the Union army.

As the Rebels regrouped and struck Bald Hill from three sides, driving the Federals onto Henry House Hill, McDowell barely had enough time to rush Union troops in front of the Stone Bridge. As night was falling, the Confederates collided with a division from Reno's corps that had arrived on Henry House Hill only moments before. The fighting ceased. Soon a heavy rain began and continued all night.

Federals elsewhere on the field withdrew quickly. Thousands surrendered, but most got away over the Stone Bridge, a few over the Sudley Springs and other fords to the north. Pope's surviving units formed a shaky line just beyond Bull Run around part of William B. Franklin's corps, which had marched up in late afternoon. Darkness had prevented a decisive Confederate victory. Lee had sent in the flank attack too late to reach Henry House Hill and the Stone Bridge before nightfall.

The next day, August 31, Jackson crossed Sudley Springs Ford and marched northwest to the Little River Turnpike (now U.S. Route 50) in an attempt to turn the Federal right and evict Pope from Centreville. The following day he struck the Federal flank guard at Ox Hill, about three miles east of Chantilly, and fought a hard battle in driving rain. The Federals lost two division commanders: Isaac I. Stevens, shot down, and Philip Kearny, killed when he rode by mistake into Confederate skirmishers. Lee sent Kearny's body back under a flag of truce.[18]

The fight at Ox Hill was enough for Pope. He withdrew into the Washington defenses. Lincoln, deeply depressed, relieved Pope and sent him to Minnesota to fight Indians. Lincoln also relieved McDowell, who never held another field command. With no other general available, he reinstated George B. McClellan.

＊   ＊   ＊

STONEWALL JACKSON HAD set up perfect conditions for Lee to destroy the army of John Pope. He had tricked Pope into attacking him at a place where the Federals' only route of retreat was over the Stone Bridge at Bull Run, entirely inadequate for an army the size of Pope's. Jackson had inflicted severe reverses on the Union army and deeply demoralized its troops. The position Jackson had selected was situated exactly where Longstreet would arrive on Pope's left flank when he came through Thoroughfare Gap. The conditions for Lee to strike swiftly around this flank and capture Henry House Hill, which dominated the Stone Bridge, were present from the moment Longstreet's corps arrived at Gainesville around noon on August 29.

But Lee had not struck on August 29. And he struck on August 30 at almost the end of the day, when there was insufficient time to seize Henry House Hill before nightfall. A plan that could have destroyed John Pope's entire army was exchanged for a modest defeat of that army. The reason was that Lee had not accepted the concept that Jackson had put forward. Jackson had proposed it on August 15, when he wanted to swing quickly around Pope's army on the Rapidan. He had actually presented it in concrete form to Lee on August 29 at Gainesville.

Jackson had shown Lee a method that could win the war in one or two engagements. But Lee had not grasped it and had not followed through. Jackson did not give up. He looked for other opportunities where Lee might accept his system. But a great opportunity had been lost at Second Manassas.

The Lost Order

IMMEDIATELY AFTER SECOND Manassas, Robert E. Lee decided to invade Maryland. His aim was to reach a decision in the war by attacking George McClellan's army on Northern soil. Stonewall Jackson, who had just demonstrated how to defeat the Union army by standing on the defensive and *then* attacking the repulsed and weakened enemy, opposed Lee's plan for a headlong assault.[1]

Although the fundamental disagreement between Lee and Jackson revolved around the question of how to fight the Union army, the initial issue with the invasion of Maryland hinged on *where* the Confederate army would go when it invaded. Lee planned to move to Frederick, Maryland, some forty miles northwest of Washington, and then march west across South Mountain, the extension into Maryland of the Blue Ridge, into the Cumberland Valley, the northern extension of the Shenandoah Valley. McClellan, compelled to follow, would be far from his main supply points. Lee planned to advance northward up the valley to Harrisburg, Pennsylvania, where he hoped to break the long railroad bridge over the Susquehanna River.[2]

Jackson opposed this move.[3] It was strategically senseless, he believed, to embark on a huge and dangerous invasion merely to interrupt Union rail connections with the West. These were not crucial to Northern prosecution of the war. The key to Union strength was the northeast. There nine-tenths of all Northern industry was located, and there the Northern population was concentrated. A deep strategic strike into the

North should go toward Baltimore and Philadelphia, not toward a railroad bridge at Harrisburg.

Even a drive on the eastern cities was not necessary, because Abraham Lincoln was supersensitive about the protection of Washington. The Confederates could threaten the capital by moving east of Frederick. At the same time a position there would menace Baltimore to the east and Philadelphia to the northeast. To Lincoln, these threats would be intolerable. Yet Lincoln was certain to require McClellan to keep his army at all times between Washington and the Confederate army. Consequently, there would be nothing to prevent the Confederates from moving toward Baltimore or Philadelphia. To stop either from taking place, McClellan would be forced to challenge Lee at the first opportunity.

General McClellan saw this obligation clearly. "As the enemy maintained the offensive and crossed the upper Potomac to threaten or invade Pennsylvania," he later wrote, "it became necessary to meet him at any cost notwithstanding the condition of the troops, to put a stop to the invasion, save Baltimore and Washington, and throw him back across the Potomac."[4]

In his official report on the campaign, filed in August 1863, McClellan wrote: "One battle lost and almost all would have been lost. Lee's army might have marched as it pleased on Washington, Baltimore, Philadelphia, or New York. It could have levied its supplies from a fertile and undevastated country, extorted tribute from wealthy and populous cities, and nowhere east of the Alleghenies was there another organized force to avert its march."[5]

Recognizing that McClellan would be forced to attack the Confederate army, Jackson proposed that the Confederate army stand somewhere in the vicinity of Frederick. Jackson's recommended dispositions are not known, but it is logical to assume that he wished the Confederate army to select a good defensive position containing at least one open flank and simply wait—just as he had done at Groveton. With the Rebels holding such a strong position, McClellan's attack would almost certainly fail, and the Confederates could then move around his flank and either destroy the Union army or throw it into retreat. This would render Washington and all the cities in the Northeast hostage to the Confederates—and almost certainly would bring peace on the South's terms.

General James Longstreet echoed Jackson's call for assuming a defensive posture, telling Lee: "General, I wish we could stand still and let the damned Yankees come to us!"[6] Nevertheless, Lee refused to accept the proposal. He wanted to attack McClellan himself, not wait for McClellan to attack him.

This determination to attack was to create problems for Lee as his army made its way into Maryland. The Union maintained garrisons at Harpers Ferry and Martinsburg, across the Potomac River in Virginia. The Federals did not evacuate the garrisons, as Lee hoped they would.

Robert Lewis Dabney, Jackson's close friend and former chief of staff, summed up the dilemma Lee faced: "His first design, of withdrawing his army in a body toward western Maryland, for the purpose of threatening Pennsylvania, and fighting McClellan upon ground of his own selection, was now beset with this difficulty: that the execution would leave the garrison of Harpers Ferry to reopen their communications with their friends, to receive an accession of strength, and to sit upon his flank, threatening his new line of supply up the Valley of Virginia. Two other plans remained; the one was to leave Harpers Ferry to itself for the present, to concentrate the whole army in a good position [east of South Mountain], and fight McClellan as he advanced." Jackson advocated exactly this, Dabney tells us. "The other [plan] was to withdraw the army west of the mountains, as at first designed [by Lee], but by different routes, embracing the reduction of Harpers Ferry by a rapid combination in this movement; and then to reassemble the whole at some favorable position in that region [of Maryland west of South Mountain] for the decisive struggle with McClellan."[7]

Lee opted for the latter plan, moving west of South Mountain and splitting up his army in order to capture Martinsburg and Harpers Ferry. With this he went against the advice of both Jackson and Longstreet, who recommended that he ignore the Union garrisons.[8] Returning to capture them would be a serious diversion of strength, his generals argued. Besides, Lee had no intention of holding the towns once they were

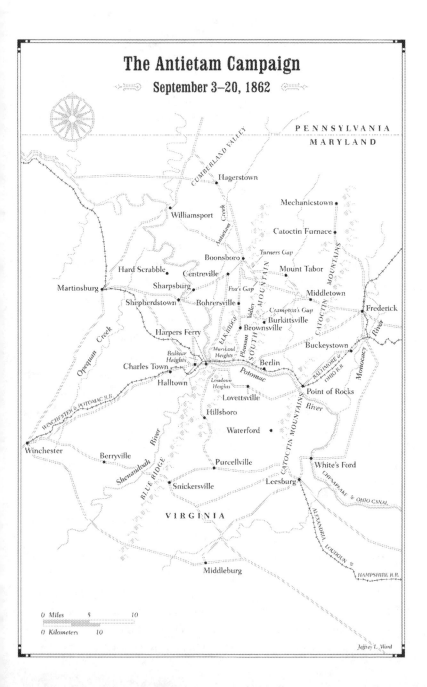

The Antietam Campaign
September 3–20, 1862

PENNSYLVANIA

MARYLAND

CUMBERLAND VALLEY

Hagerstown

Mechanicstown

Williamsport

Catoctin Furnace

Antietam Creek

Turners Gap

Boonsboro

Mount Tabor

Hard Scrabble

Centreville

CATOCTIN MOUNTAINS

SOUTH MOUNTAIN

Sharpsburg

Fox's Gap

Middletown

Martinsburg

Shepherdstown

Rohrersville

Crampton's Gap

Frederick

Pleasant Valley

Burkittsville

ELK RIDGE

Brownsville

Harpers Ferry

Maryland Heights

Buckeystown

Monocacy River

Bolivar Heights

Berlin

BALTIMORE & OHIO R.R.

Charles Town

Potomac

Opequan Creek

River

Halltown

Loudoun Heights

Point of Rocks

Lovettsville

WINCHESTER & POTOMAC R.R.

CATOCTIN MOUNTAIN

Hillsboro

Waterford

Winchester

Berryville

River

White's Ford

Purcellville

CHESAPEAKE & OHIO CANAL

BLUE RIDGE

Shenandoah

Leesburg

Snickersville

VIRGINIA

ALEXANDRIA, LOUDOUN &

Middleburg

HAMPSHIRE R.R.

0 Miles 5 10

0 Kilometers 10

Jeffrey L. Ward

captured, so there would be nothing to prevent their reoccupation after he had proceeded north.

Little recognized in analyses of the Maryland campaign is that the Confederacy's Harpers Ferry dilemma arose only because Lee insisted on moving west of Frederick in the first place. Had he remained in the vicinity of Frederick, as Jackson proposed, the garrison at Harpers Ferry would not have been an issue for the Confederates.

※   ※   ※

To Get President Davis's approval for invasion, Lee set up an elaborate deceit. He veiled his true purpose to Davis by saying he wanted to place a Confederate army in the North and then offer the Northern people peace.

"Such a proposition," he wrote Davis, "coming from us at this time, could in no way be regarded as suing for peace; but being made when it is in our power to inflict injury upon our adversary, would show conclusively to the world that our sole object is the establishment of our independence and the attainment of an honorable peace. The rejection of this offer would prove to the country that the responsibility for the continuance of the war does not rest upon us but that the party in power in the United States elect to prosecute it for purposes of their own. The proposal of peace would enable the people of the United States to determine at their coming elections whether they will support those who favor a prolongation of the war, or those who wish to bring it to a termination, which can but be productive of good to both parties without affecting the honor of either."[9]

This idea had logic behind it. Lee wanted to show the Northern people that only Lincoln and the Republicans were still striving to subjugate the South. Many Northerners now looked upon the war more as a party issue than as a national issue. But Lee also knew that Lincoln and the Republicans would never accept Southern independence.

Lee's benign and peaceable argument played directly into Davis's prejudices that the Northern people would soon tire of the war. Lee did not present his invasion as an attack on Northern factories, railroads, farms, and businesses—that is, as an attack aimed directly at the Northern people. In contrast, all of Jackson's invasion proposals had been so targeted.

But Lee's real purpose all along was to defeat McClellan in battle. Only in this way, Lee saw, could the North be truly convinced to give up the struggle. If the reverse occurred on Northern soil, it would be far more effective in changing public opinion than a reverse on Southern soil. In the wake of a Southern victory, Lee thought, many voters might turn against the Republicans in the November congressional elections. Northern governors were meeting at Altoona, Pennsylvania, in late September to challenge Lincoln's war leadership. In Britain, Foreign Secretary Lord Russell had asked Prime Minister Lord Palmerston for the cabinet to debate recognition of the Confederacy in October, a proposal designed to lead to concerted action by European powers.

Unfortunately for the South, Lee's plan called for the Confederates to assault the Union army, not for the Union army to assault the Southern army. Lee had heeded none of the lessons of the Seven Days or Second Manassas.

❋   ❋   ❋

ON THE MORNING of September 3, 1862—as soon as he was satisfied that General Pope had withdrawn to Washington—Lee turned his columns toward Leesburg and the fords over the Potomac River. The Army of Northern Virginia was on the march to Frederick, eleven miles north.

It was not a pretty army. The men were exhausted from long marches and hard battles. The horses were lank, the riders tattered, uniforms in rags, thousands without shoes.[10] Lee acknowledged the problems when he wrote Davis that "the army is not properly equipped for an invasion of an enemy's territory. It lacks much of the matériel of war, is feeble in transportation, the animals being much reduced, and the men are poorly provided with clothes, and in thousands of instances are destitute of shoes."[11]

Indeed, as the Army of Northern Virginia moved into Maryland, stragglers fell out of the ranks in record numbers—the sick, the footsore, the fainthearted, along with a few highly principled men. Mary Bedinger Mitchell, who lived at Shepherdstown, Virginia, wrote: "I saw the troops march past us every summer for four years, and I know something of the appearance of a marching army, both Union and Southern. There are always stragglers, of course, but never before or after did I see anything comparable to the demoralized state of the Confederates at this time. Never were want and exhaustion more visibly put before my eyes, and that they could march or fight at all seemed incredible."[12]

Despite the problems, Lee had concluded in his letter to Davis, "We cannot afford to be idle, and though weaker than our opponents in men and military equipments, must endeavor to harass if we cannot destroy them."[13] Fortunately for the Confederates, the men who stayed with the army remained cheerful, enthusiastic, and supremely confident. A young Maryland boy saw them march by. "They were the dirtiest men I ever saw, a most ragged, lean and hungry set of wolves. Yet there was a dash about them that the Northern men lacked. They rode like circus riders. Many of them were from the far South and spoke a dialect I could scarcely understand. They were profane beyond belief and talked incessantly."[14] Perhaps more important, the Confederates' rifles were polished and ready, and the cannons that trailed behind the artillery horses were equally clean and deadly.

McClellan got the news of the Confederate invasion from his new cavalry chief, Brigadier General Alfred Pleasonton. He shifted his headquarters to Rockville, Maryland, just northwest of Washington, on September 7, 1862. His field army numbered 85,000 men. Another 72,000 remained in the Washington defenses under Nathaniel P. Banks.

Lee halted at Frederick, seeking shoes and other supplies for his army, and hoping for an enthusiastic welcome from the citizens and many new recruits to the army. He got neither. He also hoped to confuse McClellan as to his intentions. In this he succeeded.

McClellan commenced a slow, hesitant advance on Frederick, spreading his army out on an arc twenty-five miles wide. One wing on the north, under Major General Ambrose E. Burnside, moved along the Baltimore and Ohio Railroad; another, under Major General William B. Franklin, moved along the Potomac; and a third force, under Major

General Edwin V. Sumner, advanced between them. In this fashion, McClellan shielded both Washington and Baltimore.

General Henry Halleck back in Washington was getting extremely anxious. He was afraid the move to Frederick was only a feint and that the main Confederate attack would be directed straight at Washington. He also insisted on retaining the garrisons now isolated at Harpers Ferry and at Martinsburg. The Martinsburg force was largely a 2,500-man Union garrison at Winchester that had withdrawn on the approach of the Confederates. The 10,000-man force at Harpers Ferry was under the command of Colonel Dixon S. Miles, who had commanded a division at First Manassas while drunk.

On September 9, Lee summoned Jackson to his tent and announced that he was going to divide the army into four parts. Three of them, under Jackson, were to descend on Martinsburg and Harpers Ferry, and the fourth was to move west of South Mountain at or near Boonsboro until the two places had been seized. Then the entire army was to reunite in Maryland and go on with the campaign.

Longstreet, who joined the conference, argued against dividing the army like this, as Lee's plan would leave fewer than 20,000 men in Maryland while the remainder of the Confederate army, about 25,000 men, moved back in three unsupported, separate columns on the bypassed Federal stations. But Lee overruled him. Although McClellan was advancing with vastly superior forces—numbering more than 80,000 men, Lee could guess—the Confederate commander was relying on McClellan's extreme caution and Jackson's speed and decision. The odds were greatly in favor of Jackson's seizing Martinsburg and the Ferry and getting back into Maryland before McClellan could bring himself to fall on any of the exposed Confederate forces.

Lee reasoned, too, that McClellan would not have the initiative to capture these places again once the Confederates had seized them. He thus saw the opportunity not only to capture the cut-off garrisons but also to guarantee his supply line through the Shenandoah Valley.

This thinking was characteristic of Lee. He looked for immediate, achievable goals and found it difficult to see longer-term dangers, implications, or opportunities. The momentary gain, he decided, outweighed the peril of splitting the army into fragments.

So with Special Orders No. 191, Lee directed the movement to commence the next morning, September 10.[15] Jackson was to take three divisions (12,000 men), advance on a long arc to Martinsburg, drive the Federal garrison toward Harpers Ferry (thus sealing up both the Martinsburg and the Harpers Ferry garrisons in the same bottle), and come up on Bolivar Heights, the high ground just behind the Ferry.

Lafayette McLaws, with about 9,000 men, was to descend on Maryland Heights, the dominating elevation across the Potomac from Harpers Ferry. The Confederates had to gain these heights to keep the Harpers Ferry garrison from escaping into Maryland on the pontoon bridge over the Potomac there.

John G. Walker, with his 4,000-man division, was to cross back into Virginia near the mouth of the Monocacy River, due south of Frederick, march up the south bank of the Potomac, and seize Loudoun Heights, directly east of Harpers Ferry.

Finally, Lee's orders directed D. H. Hill's division and the remainder of Longstreet's command, two divisions and an independent brigade, to march west to Boonsboro, just beyond South Mountain, while Jeb Stuart's cavalry remained to observe east of the Catoctin Mountains, the extension into Maryland of the Bull Run Mountains of Virginia.

As three parts of the army marched off on September 10 to capture Harpers Ferry, Lee changed Longstreet's orders. A report had come in that an enemy force was coming from Chambersburg, Pennsylvania—a report that turned out to be false—so Lee sent two divisions to Hagerstown to block the Federal advance and secure supplies stored there. Lee went along with Longstreet, leaving Hill's division as the only infantry force to guard Turner's Gap on the Frederick-Boonsboro Turnpike over South Mountain, as well as a northward escape route that the Harpers Ferry garrison might take.

Jackson moved with his accustomed speed.[16] He crossed the Potomac at Williamsport on September 11, spread out his divisions to intercept the Union forces at Martinsburg, and forced them to flee straight to

Harpers Ferry. Jackson's men occupied Martinsburg on the morning of September 12, and, held up briefly by the wild enthusiasm of the people, they arrived beyond Bolivar Heights at midday September 13. His men had marched more than sixty miles in three and a half days.

General Walker's division already was on the Potomac when Lee's order came through, attempting (and failing) to blow up the aqueduct of the Chesapeake and Ohio Canal over the Monocacy River, but took until September 13 to reach Loudoun Heights, only fourteen miles away. When he finally arrived, Walker found the heights unoccupied, and the next morning got five long-range Parrott rifled guns on the top.

McLaws passed over Brownsville Gap in South Mountain, some ten miles southwest of Frederick, on September 11 and advanced down the narrow Pleasant Valley between South Mountain and Elk Ridge, whose southern eminence was Maryland Heights. Learning that there was a substantial Union force on the heights, McLaws sent two brigades down the crest of Elk Ridge. He seized the heights on September 13 after scattering a green New York regiment, and the next morning he cut a road through the woods and dragged up four guns.

With cannons looking down on Harpers Ferry from both Maryland and Loudoun Heights and Jackson barring the rear at Bolivar Heights, the Federal garrison was isolated, its surrender a matter of time.

But Lee found to his shock that he had little time left. The fates, he learned, had intervened.

❋   ❋   ❋

ON SATURDAY MORNING, September 13, most of McClellan's army arrived at Frederick, three days after the Confederates had departed. The 27th Indiana bivouacked in a meadow previously occupied by D. H. Hill's division. Corporal Barton W. Mitchell noticed, among other things left behind, a bulky envelope. In it, wrapped around three cigars, was a sheet of paper, entitled "Army of Northern Virginia, Special Orders No. 191," addressed to General D. H. Hill, ending "By command of Gen. R. E. Lee" and signed "R. H. Chilton, Assist. Adj.-Gen." Mitchell had discovered a copy of the order describing in complete detail the scattering of the Confederate army.

Within minutes he got to Brigadier General Alpheus S. Williams, commanding the 12th Corps. His aide, Colonel Samuel F. Pittman, had known Chilton in the old army and verified the signature as authentic. Williams sent the order and a note asserting its veracity to General McClellan, who happened to be receiving a deputation of Frederick citizens. McClellan turned aside to read the message, became visibly elated, and shouted, "Now I know what to do!"

One of the citizens was a Southern sympathizer who recognized that McClellan had received important news relating to the Confederate movements. He rode west, found Jeb Stuart around South Mountain, and told him the story. Stuart immediately sent word to Lee at Hagerstown and D. H. Hill at Boonsboro. Thus on the night of September 13, Lee and Hill knew that McClellan had positive information as to the whereabouts of their army. As a precaution, Hill sent up Alfred H. Colquitt's brigade to guard Turner's Gap on the evening of September 13.

No one has ever learned how the lost order came to be dropped in the meadow. Hill got (and kept) a copy from Stonewall Jackson because he was under Jackson's command at the time. But Lee's headquarters also sent Hill a copy, and somehow never checked to find that it was not delivered. Hill was always suspected of receiving two copies and somehow losing one.[17]

McClellan now had the most splendid opportunity of the war to destroy the Army of Northern Virginia. He had 65,000 men at Frederick and could easily overwhelm the 30,000 men he thought Longstreet had west of South Mountain, only thirteen miles away. In fact, D. H. Hill was there with just 5,000 men. In addition, the Union general had William B. Franklin's 20,000 men only twenty miles from Maryland Heights. Franklin could march on the rear of McLaws to break the siege of Harpers Ferry and force McLaws—cut off from Jackson by the Ferry garrison—to surrender his whole command.

Yet McClellan waited an unbelievable eighteen hours—until the morning of September 14—before sending the first soldier to exploit his good fortune. This hesitation saved the Army of Northern Virginia.

Had McClellan moved on September 13, he could have sent the entire Union army—or as much as could be got together—through Crampton's

Gap over South Mountain, except for a small decoy force at Turner's Gap, which was six miles to the north. This would have given quick relief to the Harpers Ferry garrison and eliminated any chance that McLaws could join D. H. Hill, Longstreet, or Jackson. Jackson and Walker would have had great difficulty getting back across the river to join Lee. McClellan then could have turned on Hill's and Longstreet's forces north of the Potomac and destroyed them at leisure.

Instead, the Federal commander ordered the main army—30,000 men under Ambrose E. Burnside—to seize Turner's Gap. Meanwhile, he instructed Franklin to break through Crampton's Gap, put his corps into Pleasant Valley in the rear of McLaws, destroy the Confederate command, and relieve Harpers Ferry. In other words, McClellan planned to spend September 14 getting over South Mountain in preparation for a big fight beyond it on September 15.

Because of McClellan's delay in moving, Lee had just enough time. He warned McLaws that Federals would be marching on his rear, notified Jackson to conclude the siege at the Ferry, told D. H. Hill to defend Turner's Gap, and instructed Longstreet to march at daybreak on September 14 to the aid of Hill. By that morning, the Confederates were fully alert, though by no means in position to stop a resolute Union attack at Turner's Gap. Longstreet could not get to Hill until midafternoon.

❋   ❋   ❋

THERE WERE TWO battles fought on September 14 on South Mountain, one at Turner's Gap, the other at Crampton's Gap.

William R. Franklin arrived at Crampton's Gap about noon and encountered a pitifully small Confederate force—just 400 of McLaws's cavalry and infantry—behind a stone wall directly at the foot of the pass. Franklin deployed a whole division to assault this tiny force, with another division pulled up in reserve. For a while the fight looked like a stalemate, but some of Franklin's officers got exasperated with the general's timidity and ordered a charge. The thin Rebel line collapsed, and the men fled up the pass. This scared McLaws's 1,600 reserves into flight as well. All the Confederates fell in confusion back into Pleasant Valley, losing several hundred killed and wounded, plus 400 prisoners. But

Franklin, deciding that getting through the pass was enough for one day, put his corps into bivouac.

Franklin's third division reached Crampton's after the fighting ended. During the night McLaws moved up three additional brigades to strengthen the line across Pleasant Valley. Thus, the Union had 20,000 men against perhaps 5,000 Confederates. When General Franklin studied this meager line of Rebels on the morning of September 15, he concluded that it would be suicidal to attack. He somehow estimated the Confederate force to be fully as large as his own corps.

Up at Turner's Gap, D. H. Hill was slow to respond to the danger on the morning of September 14. He realized, moreover, that the Federals could approach the gap by three routes—up the main road, by way of Fox's Gap, a mile south, and by way of the Mount Tabor Church Road, coming in from the northeast through a deep gorge.

At the start of the day Hill had only Colquitt's brigade in place to defend the main road. He moved Samuel Garland's thousand-man North Carolina brigade forward, but had to commit it at once against Jacob D. Cox's Union division at Fox's Gap. Part of Cox's division reached an elevation behind the Tar Heels, sent down plunging fire, and killed Garland with the first volley. The brigade collapsed, retreated behind the mountain, gave up 200 prisoners, and was of no use for the rest of the day.

Hill stopped Cox's soldiers from moving up the ridgeline and seizing Turner's Gap only by sending down several cannons and hastily forming a guard of staff officers, couriers, teamsters, and cooks behind the guns. George B. Anderson's North Carolina brigade finally arrived and tried to regain Fox's Gap, but Cox's men pushed it back. Cox, however, didn't realize that he had actually captured the gap, and he withdrew, waiting for the rest of Jesse Reno's corps to come up.

At first Hill had no one to defend the Mount Tabor Church Road to the north. But he got a reprieve there as well. McClellan had assigned its capture to Joseph Hooker, whose corps didn't get into position until late in the afternoon. By that time Hill had gotten Robert E. Rodes's 1,200-man Alabama brigade to the ridgelines on either side of the road, and it held off Hooker's infantry for the remainder of the day.

When Hill sent Roswell R. Ripley's Georgia–North Carolina brigade to reinforce George B. Anderson's at Fox's Gap, Ripley for some unknown reason marched backward, to the rear of the mountain on the west side of the Gap. By the time he returned, the fighting was over.

At about 3:30 P.M. the first of Longstreet's men arrived, George T. "Tige" Anderson's Georgia brigade and Thomas F. Drayton's Georgia–South Carolina brigade. Hill sent them down to assist Anderson. Drayton proved incompetent, however, and his brigade buckled after a short, sharp clash with two of Reno's divisions at about 5:00 P.M. The other two brigades didn't fare well, either, and the Confederates were soon in confusion.

At this juncture, Confederate General John Bell Hood's two-brigade division, just arrived, drove down from Turner's Gap, charged through Drayton's broken troops, and forced Reno's men to the crest of the mountain. Here the Federals held as the fighting died out at nightfall, but not before General Reno fell, mortally wounded.

At Turner's Gap, meanwhile, some of Longstreet's troops kept Rodes from being overwhelmed. But the Federals seized the commanding ridgeline to the north before nightfall. In that position the Federals could shatter the Confederates in the pass the next day.

By the end of the day, therefore, it was plain to Lee that he had to abandon South Mountain and retreat, especially as Edwin Sumner's five divisions arrived in front of Turner's Gap that evening. Lee didn't learn until later in the evening that Franklin had halted in Pleasant Valley. He feared that Franklin would press on in the morning on McLaws's rear.

The battle of South Mountain cost the Federals about 2,300 men, almost all killed and wounded. The Confederates lost about 1,000 killed and wounded, but another 1,500 fell into Union hands as prisoners.

The lost order had shown McClellan where to move. Though it had not incited him to move fast, McClellan still was on the verge of interposing his army between the Confederates' three segments: Longstreet and D. H. Hill at Turner's Gap, McLaws in Pleasant Valley and Maryland Heights, and Jackson and Walker still isolated on the south side of the Potomac. Lee decided the danger was too great. The army must seek shelter across the Potomac River.

At 8:00 P.M. Lee sent off a dispatch to McLaws to abandon his positions during the night and get his two divisions behind the Potomac, across some ford. He hoped McLaws would find one downstream from the Shepherdstown ford, three miles south of Sharpsburg, for Lee wanted Hill's and Longstreet's troops to pass over it back into Virginia.

❅   ❅   ❅

MANY OBSERVERS CONCLUDED that the accident of the Union's discovering Lee's order determined the outcome of the Maryland campaign. If Lee's order had not been lost, the thinking went, the South might have succeeded.

This is not correct. If the North had not found the order, Lee still would have been defeated—for the same reason that he had been defeated at Frayser's Farm and Malvern Hill and that Union General John Pope had been defeated at Second Manassas. The defensive power of the Minié-ball rifle was causing nearly all frontal assaults to fail.

If Lee had accepted Jackson's strategic proposal to strike east of Frederick, on the other hand, the campaign would have taken an utterly different course—and would have had an almost certain likelihood of success.

❅   ❅   ❅

THE CONFEDERATES HAD fared better at Harpers Ferry than they had at South Mountain. After some difficulty communicating via flag signals, by 3:00 P.M. on September 14 the Rebel guns were firing from three directions into the town. The Federal guns, located far lower than the Confederate pieces, were virtually unable to reply. Jackson made no attempt to storm Bolivar Heights, where Dixon Miles's troops were lined up. Even so, he gained commanding ground near them, and Dorsey Pender's brigade advanced to a good assault position along the west bank of the Shenandoah River, which joined the Potomac at Harpers Ferry. Jackson ordered all artillery pieces to bear on the Federal position the next morning.

Jackson sent a message to Lee at 8:15 P.M. on September 14 telling of the progress made and that he looked "for complete success tomorrow. The advance has been directed to be resumed at dawn tomorrow."[18]

This was a strong hint that Jackson expected to capture the town on September 15.

When Lee got Jackson's message, he thought he saw a possibility of retrieving his campaign. If Jackson took the Ferry, he, McLaws, and Walker might be able to march to Sharpsburg and rendezvous with Longstreet and D. H. Hill. There was good ground for defense along Antietam Creek just east of the town. Lee immediately countermanded his order to McLaws and directed Longstreet and Hill to get their men off Turner's Gap and move the seven miles toward Sharpsburg at once, leaving the dead and seriously wounded behind.

This was the order that lost an enormous battle and perhaps the war for the Confederacy. Lee's mind was so fixed on his campaign that he did not realize that the only reasonable course open to him now was to retreat at once across the Potomac River. A hurriedly selected defensive line along Antietam Creek was not the answer. Until this night, Lee had not even considered it as a place for battle, and he knew little about the location. Boteler's Ford, just east of Shepherdstown, was the only path of retreat back to Virginia. Porter Alexander, who knew Boteler's well, described it as the worst ford on the whole river.[19] The battlefield also was in a narrow cul-de-sac, hemmed in on two sides by the Potomac River. As events were to show, it possessed no space on either flank to swing around McClellan's army if Lee was able to stop the Federal attacks and remain strong enough to launch a counterattack. At Antietam the Army of Northern Virginia sacrificed its greatest asset, its mobility. The army had to stand or die. It could do nothing else.

McClellan now held the Confederate army in a strategic vise, from which it could extricate itself only by moving back into Virginia. At Sharpsburg, even if all the dispersed elements recombined, Lee would have fewer than 40,000 men, more likely closer to 35,000, against a Federal army of 85,000. He had no possibility of winning. Even a drawn battle would have no consequence other than the Confederates' withdrawal into Virginia.[20]

Jackson had already promised a partial victory, the capture of Harpers Ferry with more than 11,000 Federal troops and much booty. If

Lee had withdrawn, the Confederacy would have achieved a signal success. More important, the North would have failed to achieve a great triumph.

But he still wanted to bring the attack to McClellan. He revealed his mind-set in a statement he made to historian William Allan on February 15, 1868, well after the war had ended: "Had McClellan continued his cautious policy for two or three days longer [after the Confederates seized Harpers Ferry], I would have had all my troops reconcentrated on the Maryland side, stragglers up, men rested, and intended then to attack McClellan, hoping the best results from the state of my troops and those of the enemy."[21]

At Sharpsburg there was no possibility of an immediate attack on McClellan. Lee's troops were too few. His initial posture had to be defensive. So Lee expected to stop McClellan's attacks and *then* go over to the offensive. But with no open flank, adopting Stonewall Jackson's defend-then-attack plan was impossible. Lee's decision to stand at Antietam demonstrated once more his focus on seeking battle to retrieve a strategic advantage when it had gone awry or he thought it had. He had shown this tendency three times in the Seven Days—at Beaver Dam Creek, Frayser's Farm, and Malvern Hill. In none of these cases was it possible to regain his advantage by resorting to the desperate, stand-up, head-on battle that he ordered. In all he suffered great losses. Now he was resolved to attempt the same frontal battle again, this time on the defensive, but hoping once more for a miracle.[22]

No one can circumvent a cardinal verity about warfare: to succeed, an army requires a leader who can see a way to victory. Without such a leader, all the valor, all the dedication, all the exertion, and all the sacrifice of an army's soldiers will go for naught.

Napoléon saw this truth most clearly. "In war," he said, "men are nothing; it is the man who is everything. The general is the head, the whole of an army. It was not the Roman army that conquered Gaul, but Caesar; it was not the Carthaginian army that made Rome tremble in her gates, but Hannibal; it was not the Macedonian army that reached the Indus, but Alexander."

The South had a general who could see, Stonewall Jackson. But Robert E. Lee was in command of the army.[23] This elemental fact decided the destiny of the South.

❋   ❋   ❋

EARLY ON SEPTEMBER 15, while Confederate guns rained shells down on Harpers Ferry, McClellan's forces advanced cautiously into Turner's Gap and discovered that the enemy had slipped away. They set off in pursuit, certain the Rebels were retreating across the Potomac. At 12:40 P.M., however, a signal observation post reported that the Confederates were forming a line of battle just beyond Antietam Creek.

The bombardment at Harpers Ferry continued until 8:00 A.M. Just as Jackson's infantry were about to storm the works on Bolivar Heights, a horseman appeared waving a white flag. Colonel Miles was surrendering. It took a while for the commanders to halt the cannon fire, however, and Miles was mortally wounded by one of the last rounds.

With orders to return immediately to Maryland, Jackson wasted no time on the surrender. He gave generous terms. All of the 11,500 Union prisoners were paroled and allowed to go home until exchanged. The officers could keep their sidearms and baggage. Jackson left A. P. Hill's division to take care of the details and immediately sent Lee news of the capture.

Late in the afternoon Jackson's two remaining divisions marched off toward Shepherdstown, halting for the night four miles from Boteler's Ford. Even Jackson called the march "severe." Behind him, Walker's division from Loudoun Heights marched, and behind Walker were McLaws's two divisions, which came down off Maryland Heights, crossed the pontoon bridge into the town, and took the road to Shepherdstown.

McClellan lost a golden opportunity to destroy Lee's small force lined up behind Antietam Creek on September 15. It numbered perhaps 16,000 men. But McClellan could not bring himself to strike fast. He decided that the next day would be soon enough to deal with this new problem.

The next morning, September 16, Jackson's divisions reached Sharpsburg and dropped down in exhaustion. Jackson, covered with dust, reported to General Lee. Soon the whole army felt more confident. Around noon one of Longstreet's soldiers came up to Henry Kyd Douglas, Jackson's aide, and asked whether Stonewall Jackson had arrived. Douglas nodded and replied: "That's he, talking to your General, 'Old Pete'—the man with the big boots on."

"Is it?" the soldier replied. "Well, bless my eyes! Thankee, Captain." Returning to his squad, the soldier exclaimed: "Boys, it's all right!"[24]

In the afternoon Walker's division arrived. McLaws's two divisions were still south of the river. The situation remained critical. McClellan had about 70,000 men forward, and even with the weary troops brought back from across the river, the Confederate army was well under half his number. Nevertheless, McClellan allowed this day, too, to pass without more than a skirmish as he moved his troops into position. He and Lee girded for battle on the morrow, September 17, 1862.

Antietam

NEARLY ALL MILITARY commentators after the Civil War criticized Robert E. Lee's decision to fight at Antietam because the Confederate army was pressed into a corner and at best could merely survive Union attacks.

Stonewall Jackson, too, was recorded initially as being opposed to fighting there. English Colonel G. F. R. Henderson in his 1898 study of Jackson relates that the Reverend Robert Lewis Dabney wrote in the first draft of his 1866 book on Jackson that Stonewall doubted the propriety of fighting the battle. Jackson's widow, Anna, insisted that Dabney submit the draft to Lee. In a letter to Mrs. Jackson on January 26, 1866, Lee disputed Dabney's statement, saying that when Jackson arrived at Sharpsburg from Harpers Ferry "and learned my reasons for offering battle, he emphatically concurred with me."

Dabney eliminated whatever negative comments he attributed to Jackson, and in the finished book he repeats Lee's language almost verbatim. Dabney was a close friend of Jackson and his chief of staff until after the Seven Days. He was generally accurate in reporting Jackson's thinking. However, Dabney was a theologian, not a soldier, and may have misinterpreted what Jackson meant.[1]

What is almost certain to have transpired is that Lee instructed Jackson—after the army had stopped the Union assaults, which Lee expected to do—to swing around the flank of the defeated Federal army and throw it into confusion and defeat. This is shown by the fact that before the end of the day of battle, September 17, 1862, Jackson, at Lee's direction, actually tried to strike around McClellan's northern flank. He was unable to do so because he had insufficient space, since the Potomac River turned up within a very short distance of the Union line there.

Lee, who decided to fight at Sharpsburg at the very last minute, had not reconnoitered the terrain. He believed the Potomac River continued due westward from Shepherdstown. He did not know that it turned north, thereby placing the entire battlefield in a narrow cul-de-sac, with the Potomac on two sides and Antietam Creek on a third. Jackson, with no opportunity of his own to study the topographical constraints, accepted Lee's judgment.

Jackson must have expressed to Dabney his great sense of disappointment that he was unable to carry out a movement around McClellan's flank. He definitely expressed this feeling to at least one other individual. This disillusionment is what Dabney conveyed in his original manuscript. Therefore, Lee in his letter to Anna Jackson reflected Jackson's opinion *before* the battle, Dabney in his first draft Jackson's opinion *after* the battle.

❅   ❅   ❅

TWO MAJOR TERRAIN features defined the battlefield. The first was Antietam Creek, about a mile east of Sharpsburg, a bold stream but fordable at places, which flowed south through a deep valley into the Potomac. The second was a broad, low ridgeline directly north of the town, along which the Hagerstown Turnpike ran.[2]

Despite the great defensive power of the Minié-ball rifle, George McClellan could have destroyed Lee's army in short order if he had taken advantage of his overwhelming numerical superiority. If, with most of his army, he had attacked down the Hagerstown Pike at the same time that he struck along Antietam Creek, Lee would have been forced to commit every one of his men. McClellan then could have sent two divisions across Snavely's Ford along the southern reaches of the creek, cut the highway to Shepherdstown, blocked the Confederates' retreat at Boteler's Ford, and crushed the Army of Northern Virginia front and rear.

This sort of tactical plan was no mystery. It was the way the Theban

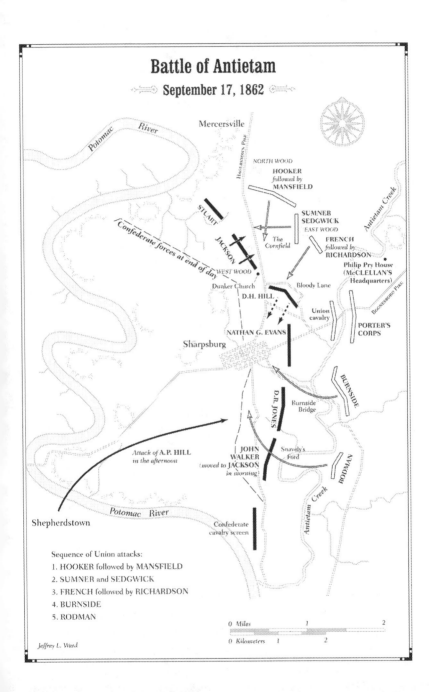

Battle of Antietam
September 17, 1862

Mercersville

Potomac River

NORTH WOOD

HOOKER
followed by
MANSFIELD

STUART

JACKSON

Confederate forces at end of day

WEST WOOD

The Cornfield

**SUMNER
SEDGWICK**
EAST WOOD

FRENCH
followed by
RICHARDSON

Antietam Creek

Dunker Church

D.H. HILL

Bloody Lane

Philip Pry House
(McCLELLAN'S
Headquarters)

Boonesboro Pike

Union
cavalry

NATHAN G. EVANS

**PORTER'S
CORPS**

Sharpsburg

BURNSIDE

D.R. JONES

Burnside
Bridge

Attack of A.P. HILL
in the afternoon

**JOHN
WALKER**
(moved to JACKSON
in morning)

Snavely's
Ford

RODMAN

Shepherdstown

Potomac River

Confederate
cavalry screen

Antietam Creek

Sequence of Union attacks:

1. HOOKER followed by MANSFIELD

2. SUMNER and SEDGWICK

3. FRENCH followed by RICHARDSON

4. BURNSIDE

5. RODMAN

0 Miles   1   2

0 Kilometers   1   2

Jeffrey L. Ward

commander Epaminondas had defeated a Spartan army at Leuctra in Greece in 371 B.C. It was how Frederick the Great of Prussia had won his battles in the eighteenth century. It had been practiced in one form or another by Napoléon Bonaparte from the Italian campaign of 1796–97 until Waterloo in 1815. Napoléon's methods especially had been discussed and argued by military writers and commanders on both sides of the Atlantic ever since Napoléon had been exiled to St. Helena.[3] But as the noted military writer Theodore Ayrault Dodge wrote in 1904, "The maxims of war are but a meaningless page to him who cannot apply them."[4]

George McClellan was one such commander. At Antietam he engaged in a battle that violated numerous maxims of war. Most important, he showed that he had no concept whatsoever of how to conduct a battle.[5] His first and most damning mistake was to keep in reserve Fitz John Porter's entire corps, plus all of his cavalry (4,200 horsemen), which he left stacked up and unused near the bridge over the Antietam on the Boonsboro Pike just east of Sharpsburg. And he scarcely used William B. Franklin's corps, which came up behind Porter.

His second fatal error was to launch attacks at three separate parts of the line, but not at the same time. His first effort went in on the north on the Hagerstown Pike, and here he compounded his error by making five separate assaults. After these disconnected attacks had run their course, he assaulted across a stone bridge over the Antietam southeast of Sharpsburg. These uncoordinated actions permitted Lee to withdraw forces from the Boonsboro Pike in the center and from the south to contain the threats to the north. And Lee was still able to get troops to deflect McClellan's fifth and last assault on the southern flank.

More than thirty years after the war, Porter Alexander was still stunned by McClellan's military incompetence. The only thing that had saved the Confederate army, he wrote, was "the Good Lord's putting it into McClellan's heart to keep Fitz John Porter's corps entirely out of the battle and Franklin's nearly out. . . . Common sense was shouting, 'Your adversary is backed against a river, with no bridge and only one ford, and that the worst one on the whole river. If you whip him now, you destroy him utterly, root and branch and bag and baggage. Not twice in a lifetime does such a chance come to any general.' "[6]

❋   ❋   ❋

MCCLELLAN OPENED THE battle with a move down the Hagerstown Pike directly toward Sharpsburg with Joseph Hooker's 1st Corps and Joseph K. F. Mansfield's 12th Corps. He had, however, moved both corps into position the day before, thereby signaling where the attack was coming from. This gave Stonewall Jackson, posted on the north, ample time to move his corps into blocking position during the night of September 16–17.[7]

At 5:00 A.M. on September 17 Jackson's force stood mainly west of the Hagerstown Pike facing the Union attack. John R. Jones's Confederate division rested in and near the West Wood, a block of mature hardwood to the west of the pike and extending northward from the Dunker church, a whitewashed brick building on the pike a mile north of Sharpsburg. Alexander R. Lawton's division was in the open partly to the east of the pike, and just south of a thirty-acre unharvested field of Indian corn. The tall stalks of this maize could hide the advance of the enemy. On the eastern edge of this field, soon to be known as the Cornfield, where much of the fighting was concentrated, was another long band of mature hardwood, known as the East Wood.

The Federal soldiers went bravely into battle but were unable to break through because the assaults went in separately and individually. Jackson stopped the Union effort with a counterattack using Tige Anderson's Georgia brigade from the Boonsboro Pike and John G. Walker's entire division from the lower Antietam, all hurriedly sent by Lee to Jackson. Lee also sent up Lafayette McLaws's division, just arriving, to reinforce Jackson.

Since Lee had stripped bare other portions of the Confederate line to support Jackson, a determined Federal attack at any point could break the Rebel line. Lee had only Richard Anderson's division as reserve. A. P. Hill, though marching hard from Harpers Ferry, was still miles away.

About this time, Jackson's medical officer, Hunter McGuire, appalled by the number of wounded and assuming the day was going badly, rode to the West Wood to discuss whether to transfer wounded across the Potomac. He found Jackson sitting quietly on his horse, Little Sorrel, behind the line of battle. McGuire was not reassured by what he saw at the West Wood. Men in many places were lying at intervals of several yards. Over the Cornfield the strength of the Federal masses was terribly apparent. "Can our line hold against another attack?" McGuire asked anxiously. Jackson told him not to worry. "Dr. McGuire, they have done their worst," he said. "There is now no danger of the line being broken."

At 9:10 A.M. McClellan learned that Franklin's corps was approaching. He finally decided it was safe to order Ambrose Burnside's four-division 9th Corps to assault the Confederate right, or southern, flank. He also released Israel Richardson's division, which he had been holding in reserve, and it started toward the East Wood.

It took nearly an hour to deliver the movement order to Burnside, and longer for him to go into action. In the meantime, William French, whose 5,700-man division was standing alone in the East Wood, decided to advance southwest. This threw French directly against 2,500 men from D. H. Hill's division who were facing northeast along a zigzagging sunken road (the "Bloody Lane") beginning six hundred yards south and east of the Dunker church.

French sent in his brigades one at a time headlong against the sunken road, and each brigade was shattered in its turn. In the space of minutes 1,750 men were killed or wounded, and most of the remainder of French's division had fled in panic to the rear.

Now, after more than two Federal corps had been destroyed, Richardson's division of 4,000 men arrived. It, too, aimed toward the new vortex of the battle, the Bloody Lane. D. H. Hill's men were exhilarated but exhausted, their numbers dwindling. Losses were especially great among officers.

Longstreet and Jackson desperately fed every man and cannon they could find into the struggle. Lee also committed his last force, Richard Anderson's 3,400-man division. As it marched over rising ground south of Hill's position, however, Federal Parrott guns east of the Antietam found the range and scored a number of casualties, among them Anderson himself, who was severely wounded. Brigadier General Roger A.

Pryor, a former congressman, took over for Anderson but was unable to handle command.

Richardson advanced to the east of French's division, whose remaining men huddled under the crest of the hill. Richardson's lead brigade was likewise shattered, losing two-thirds of its strength in a couple of minutes. But his second brigade went in a little to the left of the first brigade and happened to catch the Confederates in disorder. Some men from Anderson's ill-led division, who had moved into the sunken lane, panicked and fled.

Almost at the same moment, on the left side of the sunken road, Lieutenant Colonel J. N. Lightfoot, commanding the 6th Alabama, asked his brigade commander, Robert E. Rodes, for permission to withdraw from a small exposed salient. Rodes agreed, but Lightfoot misinterpreted his orders and told the commander of the adjacent regiment that the withdrawal order applied to the whole brigade. This weakened the entire Confederate line along the sunken road and sent men fleeing to the rear. Rodes desperately rallied a few men, but the defection blew a wide hole in the center of the Confederate line.

Richardson's men, with some of French's soldiers coming along, rushed gleefully into this void. They quickly launched devastating enfilade fire against the few Confederates still grimly holding parts of the sunken road, killing many Rebels and forcing the rest to flee.

Richardson's breakthrough was clearly visible to McClellan at his observation post east of Antietam Creek. Officers there now agreed that the time was perfect to launch Fitz John Porter's corps at the center of the Confederate line at the bridge on Boonsboro Pike. But McClellan refused. He held Porter in place and sent no help to Richardson.

Thus McClellan did not exploit the one true breakthrough his troops had achieved. Massed Confederate artillery south of the sunken road stopped Richardson's advance. Richardson, getting no support, ordered his men to safety behind the ridgeline just north of the sunken road. Richardson was wounded by a Confederate shell. (He would die six weeks later.) McClellan replaced Richardson with Winfield Scott Hancock and told Hancock to stay where he was.

❖   ❖   ❖

As THE FIGHTING was slowing on the north, it commenced on the south, demonstrating the uncoordinated nature of McClellan's battle plan. On this flank Burnside sent one charge after another to force passage of the Rohrbach Bridge—to go down in history as the Burnside Bridge. He was sending his men directly into the barrels of 400 Georgians under Colonel Robert L. Benning, who were protected by a stone wall and rock quarry, with twelve cannons in close support and two more batteries posted on the plateau next to Sharpsburg.

At the cost of 500 Union soldiers Burnside at last succeeded. In the late afternoon he formed his corps on the western side of Antietam Creek and began a march on the town of Sharpsburg and the road leading south to Boteler's Ford.

Just at this moment, however, A. P. Hill's division, completing its forced march from Harpers Ferry, arrived on Burnside's flank, shattered the advance, and drove Burnside's entire corps back up against Antietam Creek.

While Hill was repelling the Federals on the south, Lee and Jackson were seeking to attack on the north. Although Jackson had no more than 5,000 men, including cavalry, he examined the northern front, hoping to swing around the northern flank. He discovered at least four Union brigades and many cannons in the way of the only route he could take, since the Potomac prevented a wide flanking movement.

Jackson's men had to pass through a corridor less than a mile wide between the Potomac and Union General George Meade's position just above the North Wood. With Jeb Stuart's twenty-one guns, most of them light smoothbores, Jackson tried to silence Meade's thirty-four rifled guns, which were ranked hub to hub on high ground. But Meade's artillery overwhelmed Stuart's pieces in about fifteen minutes. Stuart and Jackson sadly concluded that it was impossible to turn McClellan's flank, and the men and guns returned to their starting places.[8]

Shortly afterward, General Walker asked Jackson about the failed counterstroke. Jackson told him it could not be carried out because Stuart had found that the Federal right was securely posted on the Potomac and there was no space for a wide turning movement. Jackson said he was surprised, because he had supposed the river to be much farther away. "It is a great pity," Jackson said, "we should have driven McClellan into the Potomac."[9]

It was now nightfall. The fighting ceased all across the line. It had been the bloodiest day in American history. The Army of the Potomac had suffered 2,108 dead, 9,450 wounded, and 753 missing, many of them dead as well—a total of 12,401, one-quarter of the men who had gone into action. The Army of Northern Virginia had lost about 1,500 dead, 7,800 wounded, and 1,000 missing (most dead)—a total of 10,300, 31 percent of those on the firing lines. The combined American total was 22,700.

McClellan had committed only 50,000 men. Over a third of his army had not fired a shot. Even so, his determined and brave soldiers came close to cracking the Confederate line and bringing Lee's army to disaster.

The Army of Northern Virginia now entered a new and bitter era. It had been deeply, permanently wounded. The aura of inevitable victory no longer enveloped it. It had fought McClellan to a standstill, but it had not achieved victory. Soon it would have to retreat. This, to the world, would spell defeat. Abraham Lincoln was waiting to proclaim emancipation of the slaves and transformation of the war into a crusade. This proclamation would guarantee the persistence of the Northern people in the struggle and end any hope that Britain and France would recognize the Confederacy—exactly what Jefferson Davis had been counting on since the war's beginning.

❊   ❊   ❊

WHILE THESE EVENTS were taking place in the East, the Confederacy's fortunes in the West were declining even more rapidly. The primary culprit was General Braxton Bragg, whom President Davis had appointed to command on June 27, 1862, and who showed himself to be incompetent.

After the Confederacy's defeat at Shiloh on April 6–7, 1862, the Southern army had withdrawn into northern Mississippi. Beginning in June, Union Major General Don Carlos Buell commenced a long, slow march eastward to capture Chattanooga, Tennessee. Bragg intercepted this movement by transferring troops by rail from Tupelo, Mississippi, to Chattanooga. In early August 1862 he and the Confederate commander in east Tennessee, Major General E. Kirby Smith, embarked on a campaign to invade Kentucky. Smith moved from Knoxville into eastern Kentucky and, after defeating a Union force around Richmond, Kentucky, reached Lexington and Frankfort, the capital. Buell, to protect Nashville, withdrew to Decherd, Tennessee, fifty miles west of Chattanooga.

By September 14, Buell had withdrawn to Bowling Green, Kentucky, while Bragg had slipped past him and occupied Munfordville, on the Green River forty miles to the northeast. Bragg was now blocking the road Buell had to travel to reach the main Union base of Louisville.

But instead of standing at Munfordville and calling on Kirby Smith's troops at Lexington to reinforce him, Bragg withdrew without a fight northeastward to Bardstown, Kentucky. This was a disastrous decision. Buell quickly took the clear path to Louisville that the Confederate general had opened. Had Bragg stayed at Munfordville, he could have stopped Buell, whose supplies were running out. At best Buell would have had to retreat back to Nashville, and at worst he would have been forced to surrender. Bragg could easily have captured Louisville, which was practically undefended.

This would have changed the whole nature of the war. At Louisville the Confederates could have broken all existing Union supply lines to the armies in the West, blocked boat traffic on the Ohio River, and, by means of cavalry strikes into southern Indiana and Illinois, slowed rail connections to St. Louis. This would have made it extremely difficult for the North to set up new supply routes to provision its troops around Memphis, Nashville, and Corinth, Mississippi.

But Bragg refused to seize the opportunity. Even stranger was his decision after the reinforced and resupplied Buell pressed battle on him on October 8, 1862, at Perryville, about forty miles south of Frankfort. Because of atrocious planning, Bragg got only 16,000 men on the field—about a third of his and Smith's forces. Even so, the Confederates drove back the Union army and won the day. Yet Bragg immediately decided to retreat back into Tennessee. By the end of the year 1862, Bragg was nervously resting at Murfreesboro, about eighty miles northwest of Chattanooga, the initiative having passed entirely into Union hands.

Because of Bragg's command failures, the North had essentially won the war in the West. It was going to take another year to seize Vicksburg and open the Mississippi to traffic, and to capture Chattanooga and open the road to Atlanta. But, with President Davis unwilling to remove his old friend, the handwriting was on the wall.

9

Fredericksburg

O
N THE MORNING of September 18, 1862, Lee and Jackson reluctantly concluded that they could not launch a successful attack around McClellan's northern flank because there was insufficient space between his lines and the Potomac. The only choice left was to retire across the river. Lee held the army in place all day to permit evacuation of the wounded. But after midnight and on into the morning of September 19 the Army of Northern Virginia crossed Boteler's Ford east of Shepherdstown and bivouacked below the southern shore.

Lee left his reserve artillery and a small infantry detachment to guard the ford. Union General Fitz John Porter pushed a small force across that seized four guns. But Stonewall Jackson sent in A. P. Hill's division on September 20, driving the Federals back across the river.

That ended the Maryland campaign. But its consequences reverberated to the end of the war and beyond. On September 22 Abraham Lincoln issued the Emancipation Proclamation, effective January 1, 1863. It was a revolutionary presidential edict that struck down constitutional protections for slavery and changed the war from a conflict to preserve the Union into a crusade to free the slaves. The edict did *not* end slavery in the Northern border states that had not seceded, only in the eleven Confederate states that had withdrawn.[1]

Lincoln wanted McClellan to go after Lee's army at once, but McClellan commenced his usual procrastination, remaining in place and demanding that his army be refurbished before he could move. This angered Lincoln. Two other suspicions arose in the president's mind—that McClellan would be the Democratic peace candidate in the presidential election of 1864 and that he was deliberately avoiding destruction of the Southern army to prepare for a compromise peace that would save slavery.[2]

On October 6, 1862, Lincoln gave McClellan a peremptory order to "cross the Potomac and give battle to the enemy." Once more Lincoln tried to be a strategist, advising McClellan to take the "interior line," keeping between the Rebels and Washington and promising him 30,000 more men if he did so instead of advancing up the Shenandoah Valley, as McClellan preferred.[3]

The Army of Northern Virginia, meanwhile, underwent a remarkable recovery. Resting in ideal weather in and around Winchester, and receiving ample food, the army acquired much-needed clothing, shoes, and blankets. The few men who still were armed with smoothbore muskets exchanged them for rifles picked up on the battlefield or captured at Harpers Ferry. The army more than doubled in size as stragglers returned and a constant, if small, stream of conscripts arrived. On October 10, the army numbered 64,273; by November 20, 76,472.[4] Lee also superintended the reorganization of the army formally into corps, one led by Jackson, the other by Longstreet. He promoted both officers to lieutenant general.[5]

For a while Lee entertained the idea of advancing again into Maryland, but eventually he gave it up as too hazardous. Even so, he wanted to know McClellan's plans and on October 10 sent Jeb Stuart north with 1,800 horsemen. The target was the railroad bridge over Conococheague Creek at Chambersburg, Pennsylvania, to cut McClellan's rail supply line from Harrisburg to Hagerstown.

Stuart crossed near Williamsport and reached Chambersburg without incident. Though he couldn't destroy the bridge, which was made of iron, his troopers rounded up 1,200 sorely needed horses. With large bodies of Federal cavalry and infantry trying to head him off, Stuart decided it was safer, and far more dramatic, to make an entire circuit of McClellan's army. He returned by way of Emmitsburg and Hyattstown, bluffed half a regiment to retreat from its position guarding White's Ford, four miles north of Leesburg, and returned safely to Virginia on October 12, having completed a 126-mile circuit, losing one man wounded and two captured.

This second ride around McClellan's army, repeating the one he had made in June on the peninsula, damaged McClellan's reputation even further, especially with Lincoln.

Despite Lincoln's order, it was October 26 before McClellan began a leisurely crossing of the Potomac at Harpers Ferry and five miles east at Berlin (present-day Brunswick). It took eight days. The leading elements moved east of the Blue Ridge and advanced about twenty miles into Virginia. The next stage was a move on Warrenton, another twenty miles south. There McClellan could exchange his difficult supply line through Harpers Ferry with the safer and easier Orange and Alexandria and Manassas Gap railroads. By November 4 he had secured the new rail connections, and substantial reinforcements had arrived from Washington.

That same day congressional elections were held in the North. Seven states that had voted Republican in 1860 went Democratic in 1862, though the Republicans retained control of both houses of Congress. The results could have been much worse for Lincoln and the Republicans, especially if, on election day, a Confederate army had been ranging through Pennsylvania and Maryland instead of standing defeated back in Virginia.

Nevertheless, Lincoln had had enough of McClellan. Two days after the elections he deposed the general, replacing him on November 7 with Major General Ambrose E. Burnside. An officer with little imagination and less energy, Burnside honestly protested that he was not qualified for high command; it was an opinion shared by nearly all the senior officers in the Army of the Potomac. He took the job only because Lincoln said he would otherwise appoint Joseph Hooker, whom Burnside considered to be a devious conniver who had conspired to get an independent command at Antietam.[6]

Burnside knew that Lincoln expected quick action. A directive accompanied Burnside's appointment ordering him to report to Lincoln what he expected to accomplish. Two days later Burnside recommended that the army move to Fredericksburg, Virginia, where it could get its supplies by ship through Aquia Creek, twelve miles away on the Potomac.

After seizing Fredericksburg, Burnside said, the army would push toward Richmond, fifty-five miles away.[7]

A far better target would have been Lee's army itself. Lee had already divided the Army of Northern Virginia, once he was certain that the Federals were moving east of the Blue Ridge Mountains. He had taken Longstreet's corps and marched rapidly to Culpeper, ahead of the slow Federal advance, to interpose his army between the Union force and the Virginia Central Railroad, whose closest point was at Gordonsville, about thirty miles south. Lee had left Jackson in the Shenandoah Valley for the chance to sally on the flank and rear of the Federal army. Thus the two Rebel corps were two days' march apart, separated by the Blue Ridge. Burnside now had 119,000 men and 374 guns only a day's march away from Longstreet's corps at Culpeper. But striking at Lee's army did not occur to Burnside, or to any other Federal in authority.

General Henry Halleck, in Washington, preferred a repetition of John Pope's plan of moving toward Gordonsville. But Lincoln said he would agree with Burnside's proposal provided the general moved fast, crossed the Rappahannock River upstream from Fredericksburg, marched down the south bank, captured the town from the flank and rear, and seized the heights parallel to the river about a mile south of the town. These heights were the key to the defense of Fredericksburg, and flanking them was far preferable to assaulting them head-on.

Burnside did precisely the opposite. He approached the town from Falmouth, directly across the river.

General Edwin V. Sumner's Union advance force reached Falmouth on November 17, while there was only a tiny Confederate observation force at Fredericksburg. Sumner proposed to cross the river at once and seize the town and the heights behind. But Burnside said no, he was to await the arrival of pontoon bridges. These took until November 25 to arrive.

Longstreet got there first. Lee ordered part of his corps to Fredericksburg (thirty-five miles from Culpeper) on November 18 and the remainder the next day. Longstreet took up positions on the heights south of the town, at last awakening Burnside to the difficulties of assaulting them. Alarmed, the Union general reconnoitered possible flank movements. Above the town the country was hilly and wooded and the river narrow, with several fords. Here a surprise crossing would have been easy. Instead, Burnside looked only downstream, where the river was tidal and broad. He began preparations to cross at Skinker's Neck, twelve miles below the town, and also investigated Port Royal, eighteen miles below Fredericksburg, where the river was a thousand feet wide.

※   ※   ※

STONEWALL JACKSON, seeing that Lee was girding to defend Fredericksburg, proposed an entirely different strategy.

He certainly recognized that the heights just south of Fredericksburg offered an easy defensive position, especially since there was open ground 1,000 to 1,800 yards wide between the heights and the river. Over such ground a frontal attack by the Federals had little chance of success. Thus a Confederate victory was almost guaranteed.

Nevertheless, Jackson deeply opposed fighting the Union army there. If the Union should cross at Fredericksburg, both its flanks would be protected by the river. And its numerous batteries, arrayed on Stafford Heights, an elevation directly across the river from the town, would command the length and the breadth of the battlefield, making a counterstroke difficult and pursuit impossible.[8] To await attack, moreover, would surrender the initiative to the Federals, for Burnside could choose his own time and place to attack. Lastly, it would be impracticable to maneuver against Burnside's most vulnerable point, his supply line. If defeated and counterattacked, Burnside could easily withdraw the dozen miles back to Aquia Creek before the Rebels could cut the line and isolate his army.

The North Anna River, about twenty-five miles south, promised far greater results, in Jackson's view. The Federals, advancing from Fredericksburg, would expose their right flank and their communications for the entire distance from Aquia Creek to the North Anna—about thirty-seven miles. If they attacked on the North Anna and were defeated, as Jackson was confident would be the case, they would have to retreat all the way back to Aquia Creek, and the Confederates would have a great opportunity to block this retreat and destroy the entire Union army.

"I am opposed to fighting on the Rappahannock," he said to General D. H. Hill. "We will whip the enemy, but gain no fruits of victory. I have advised the line of the North Anna, but have been overruled."[9]

Three times since mid-August Jackson had proposed a plan to annihilate the Northern army. Lee had delayed implementing his first proposal, to crush John Pope against the Rapidan, until it was too late. At Second Manassas, Jackson had drawn Pope into attacking with inadequate routes of retreat and an undefended flank, but Lee again had waited until it was too late. In the Maryland campaign Jackson wanted to place McClellan on the horns of a dilemma: to attack the Confederate army and lose or to give up Philadelphia and possibly Baltimore and still lose. Lee instead followed his own plan, which resulted in the disaster of Antietam.

Now for the fourth time Lee rejected Jackson's strategy to win the war in a swift campaign by eradicating the Northern army. Lee decided to stay at Fredericksburg. He was virtually assured of victory here, but nothing would be gained from it. The only reason Lee ever cited for his decision was that he wanted to deny to the enemy the territory between the Rappahannock and the North Anna.

Jackson protested, but to no avail. He loyally moved his corps to Fredericksburg on November 29.[10]

※   ※   ※

LEE SENT JACKSON'S corps to cover both Skinker's Neck and Port Royal. When Burnside discovered this movement, he conceived a new idea. He would cross at Fredericksburg very quickly and unexpectedly, and drive around Longstreet's right, or east, at Hamilton's Crossing, at the end of Prospect Hill, five miles southeast of Fredericksburg, the most easterly portion of the line of heights south of the town.[11] Here both the railroad and the Richmond Stage Road turned south and skirted the edge of Prospect Hill. If Burnside could get around this flank, he would interpose his army between Longstreet and Jackson at Skinker's Neck and Port Royal.

This plan would work only with swift and resolute action. Burnside was obliged to strike before Jackson could bring his corps back to Fredericksburg. Burnside had to force passage of the Rappahannock, march

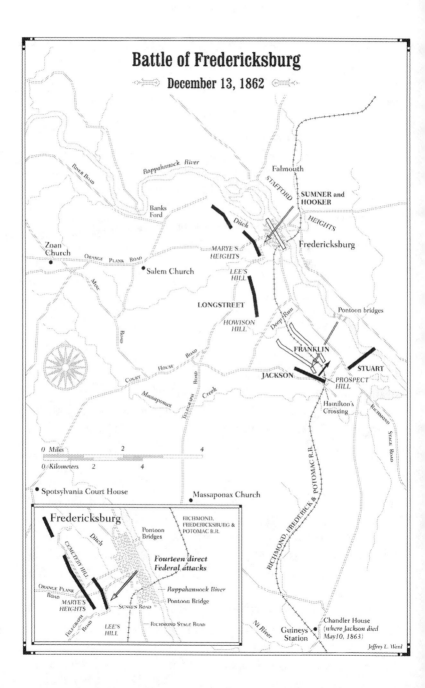

Battle of Fredericksburg
December 13, 1862

River Road

Rappahannock River

Falmouth

STAFFORD

SUMNER and HOOKER

Banks
Ford

Ditch

HEIGHTS

Zoan
Church

ORANGE PLANK ROAD

*MARYE'S
HEIGHTS*

Fredericksburg

Salem Church

*LEE'S
HILL*

Mine

LONGSTREET

*HOWISON
HILL*

Deep Run

Pontoon bridges

Road

Road

HOUSE

FRANKLIN

ROAD

COURT

JACKSON

STUART

*PROSPECT
HILL*

Massaponax

Creek

TELEGRAPH

Hamilton's
Crossing

RICHMOND

RICHMOND STAGE ROAD

0 Miles   2   4

0 Kilometers   2   4

• Spotsylvania Court House

• Massaponax Church

RICHMOND, FREDERICK & POTOMAC R.R.

Fredericksburg

CEMETERY HILL

Ditch

Pontoon
Bridges

RICHMOND,
FREDERICKSBURG &
POTOMAC R.R.

***Fourteen direct
Federal attacks***

*ORANGE PLANK
ROAD*

*MARYE'S
HEIGHTS*

Rappahannock River

Pontoon Bridge

Sunken Road

TELEGRAPH ROAD

*LEE'S
HILL*

RICHMOND STAGE ROAD

Ni River

Guiney's
Station

Chandler House
(where Jackson died
May 10, 1863)

Jeffrey L. Ward

along the narrow plain between the river and the heights, directly under Longstreet's guns, then swing around Longstreet's flank and onto his rear. He had confidence in the plan. Pointing out the heights to William Farrar Smith, commander of the 6th Corps, in late November, Burnside said: "I expect to cross and occupy the hills before Lee can bring anything serious to meet me."[12]

Once in a while an ordinary general sees how he can accomplish a victory by a wholly unexpected maneuver, as Burnside did in this instance. But only masters of war normally have the boldness and strength of character to suppress doubts, ignore timid counsel, concentrate their forces, and hit the enemy's decisive point quickly and forcefully. The mediocre general usually acts too slowly and too irresolutely. As a result, the blow normally does not achieve the results expected. When the enemy responds swiftly and the situation changes before his eyes, the mediocre general usually tries to prevent disaster by striking directly at the enemy, forgetting that the power of his idea had been to *avoid* the enemy's strength.

This is what happened to Burnside. His idea became a trap of his own making. He moved far too slowly, tipped his hand, found the enemy in strength in front of him, lost his nerve and his sense of purpose, and struck headlong against the most powerful and prominent positions of the enemy, seeking to destroy by brute force what he had been unable to defeat by guile.

※   ※   ※

ALTHOUGH DETERRED FOR hours by Confederate snipers, Burnside got two pontoon bridges across the river at the upper end of town, another at the lower end, and three a mile below the town on December 11.

To flush out the Rebel riflemen, Burnside ordered all the artillery on Stafford Heights to shell the town. This cannonade created havoc but missed most of the Rebel riflemen and also gave Burnside's intentions away. Lee now knew for certain that Burnside was planning his major assault at Fredericksburg, and sent word to Jackson to bring up his forces from Skinker's Neck and Port Royal.

Burnside now compounded his error. Although the only merit of the plan was to catch Lee's army dispersed, he gave the Confederates time to unite by spending all of December 12 crossing much of his army over the bridges and assembling them on the southern shore. By noon on the 12th, Jackson already had moved up the divisions of A. P. Hill and W. B. Taliaferro and hidden them in the woods of Prospect Hill, four miles southeast of Fredericksburg. This formed the extreme right of the Confederate position, its easternmost terminus just above Hamilton's Crossing and the valley of Massaponax Creek, which flowed into the Rappahannock two miles to the east. D. H. Hill's division from Skinker's Neck and Jubal Early's division from Port Royal were on a hard march for the heights. They arrived on the morning of December 13, forming a reserve on Prospect Hill, and thus uniting the entire Confederate army of 78,000 men.

On December 12, General Edwin Sumner brought six Federal divisions over the upper bridges and occupied the town. He faced part of Longstreet's corps that was spread out on and under Marye's Heights, which rose 130 feet just behind the town.

Major General William B. Franklin crossed the lower bridges with six more Union divisions. Franklin faced Jackson's corps, which was positioned for a little over a mile on Prospect Hill. Two additional Union divisions remained just north of the lower bridges, ready to cross and support Franklin. He had nearly 60,000 men and ample guns available against Jackson's 30,000 infantry.

Burnside's delays had destroyed any possibility of a swing around Lee's eastern flank. Sensing this, Burnside elected to win by direct assaults. He ordered Franklin to send "a division, at least" to seize Prospect Hill. After this attack had started, and then only on Burnside's order, was Sumner to direct "a division or more" to move up the Orange Plank Road (now Virginia Route 3) and along the Telegraph Road "with a view to seizing the heights to the rear of the town." The Plank Road was two blocks (an eighth of a mile) west of the Telegraph Road, which started at Hanover Street at the edge of town, ran parallel with the Plank Road for a quarter mile, then veered off to the east, skirted the base of Marye's Heights, and turned south through the valley of Hazel Run.

Around Marye's Heights, the Federal commanders made no effort to advance directly up the Plank Road—that is, on the flank of the heights. They also attempted no assault farther west, where there were only a few Confederate troops. Instead, the assaults went in entirely against the steepest part of the heights, directly against the Telegraph Road.[13]

The Federal battle was lost before it began. Burnside made two isolated attacks on the left and right, neither of which could be decisive and neither of which could support the other. In addition, he set out to assault the two points on Lee's line that were best prepared to receive him.

The first point, Jackson's line, was protected by woods that prevented accurate enemy artillery fire and allowed the Confederate infantry to hide. Furthermore, the line was so short that Rebels could be concentrated in great numbers. Indeed, Jackson did not have enough space along the line to deploy all of his infantry. Jackson deployed only A. P. Hill's division on the front. His three other divisions remained in support. Protecting Jackson's line were fourteen cannons emplaced just above Hamilton's Crossing, twelve guns to the left that were advanced north of the railroad line that ran at the base of the ridgeline, and twenty-one more guns emplaced on the hill about 200 yards behind.

The point that Burnside had directed Sumner's troops to assault was equally impregnable. Telegraph Road, along the base of Marye's Heights, was sunken a couple of feet below grade and flanked by a stone wall, which, when banked with dirt thrown up around it, formed a perfect defensive parapet impervious to rifle fire. About 2,000 Confederates were defending this "Sunken Road" over about a quarter of a mile. In support on Marye's Heights were about 7,000 more Rebel infantry, plus nine cannons in easy canister range of any enemy approaching the Sunken Road. On the heights to the left, or west, of the Plank Road were eight more Confederate cannons. Eight additional guns were on Lee's Hill (originally Telegraph Hill), a 210-foot elevation a mile to the east where General Lee made his headquarters. These guns could enfilade any advance against Marye's Heights.

The Federals did have one opportunity, however: over on Jackson's front line, A. P. Hill had deployed his six brigades well except for the single glaring error of leaving a gap of about five hundred yards between the brigades of James J. Archer and James H. Lane. This was the only part of Jackson's line where the woods extended beyond the railroad line, and the marshy, overgrown woodland offered a covered approach route for the enemy.

In the plain between Prospect Hill and the river, Franklin lined up George G. Meade's division on his left, with John Gibbon's division supporting it on the right rear. Abner Doubleday's division deployed on the extreme left to defend against Stuart's cavalry.

Meade's brigades were held up for a time by Stuart's artillery chief, Major John Pelham, who advanced two cannons on the flank of Meade's brigades and opened fire. A dozen Federal guns tried to silence Pelham's guns, and did disable one, but "the gallant Pelham," as Lee described him, held up Meade's advance for an hour before Jackson ordered him to withdraw.

After a heavy Federal cannonade, Meade's division, supported by Gibbon on the right, advanced. Meade's force, discovering the wooded marshy area to their front to be free of Confederate missiles and hidden from view of Rebel infantry, gratefully pressed into this vacuum. Despite the soft footing, they advanced entirely through the woods and fell on Archer's left flank and Lane's right, capturing about 300 prisoners, routing two of Archer's regiments, and forcing Lane's brigade back into the woods.

Meade's center brigade, meeting no resistance, bore down on Maxcy Gregg's South Carolina brigade in reserve on a road just behind the lines. Gregg's men grasped their weapons and began firing, but Gregg thought the force was Confederate and rushed forward to stop the firing. Almost instantly he was mortally wounded. One of Gregg's regiments collapsed, but the rest held firm.

Meade's brigade was now in deep trouble. It had penetrated far into the center of the Confederate position without support and had already fallen into confusion because of the resistance of Gregg's brigade. Jackson ordered Early and Taliaferro to advance and clear the front. They crashed into the Federals, driving Meade's entire division in panic and disorder out of the woods, beyond the railroad line, and out into the plain.

Gibbon's division likewise fell back in the chaos that ensued. The Confederates were carried along for a distance by the passion of battle but soon moved back to the railroad.

Meanwhile, Burnside had ordered Sumner to attack Marye's Heights. Here a violent battle was raging. There was never any possibility that this attack could support Franklin's effort, since the two assaults were four miles apart. Burnside thus fought two separate battles on December 13, 1862, neither related to the other.

Like Franklin's move against Jackson, Sumner's assault also was on a narrow front. In the space of four hours, Burnside sent fourteen separate attacks against the Sunken Road. All failed, with appalling casualties. It was an insane waste. The results of each assault were a foregone conclusion, but Burnside obstinately insisted that the frontal attacks continue.

While one of the attacks was collapsing against the stone wall, General Lee put his hand on Longstreet's arm and said, "It is well that war is so terrible, or we would grow too fond of it."[14]

       ❈   ❈   ❈

THE HORRIBLE DAY of battle at last ended. Only two parts of the entire Confederate line had been tested. Large portions had stood idle all day. Most of the Confederate officers believed that Burnside would renew the battle the next day. Burnside indeed intended to do so. His proposal for December 14 reveals his incompetence as a commander. The only plan he could think of was to form the entire three-division 9th Corps into a column of regiments and lead it in person upon Marye's Heights—thus repeating the maneuver that had failed utterly on December 13. Sumner, Hooker, and Franklin talked him out of this scheme.

The cost of the battle was staggering. The Federals lost 12,647 men, most killed and wounded, the Confederates 5,309. The disparity was greatest at Marye's Heights, with the Federals suffering five times as many losses as the Confederates.[15]

The two armies exchanged some desultory fire on December 14. On December 15 Hooker and Franklin applied to Burnside for a truce to permit removal of the Federal wounded, unattended between the lines. Burnside refused, and the wounded around Marye's Heights remained.

On Franklin's front, however, Jackson honored an informal cease-fire and the Union medical attendants tended to the surviving wounded and buried the dead. The night of December 15 was dark and rainy. Burnside used it to withdraw his entire army north of the river.

Stonewall Jackson had been right. The Confederates had whipped the Federal army but gained no advantage. The Federals had not been forced into disorder and retreat. They retained the initiative and easily could replace their losses. Jackson's hope for a great and decisive Confederate victory remained as elusive as ever.

Chancellorsville

THE BATTLE OF FREDERICKSBURG aroused unrealistic hopes in Confederate government circles and among the Southern populace. James Longstreet reported that authorities in Richmond assured General Lee the war was virtually over. Gold advanced to $200 an ounce in New York, and senior officials said that in thirty or forty days the Confederacy would be recognized and peace proclaimed. Lee did not share this optimism.[1]

The Southern civilian leadership refused to look soberly at the peril facing the Confederacy. President Davis continued to hope the Northern people would tire of the war, even after Abraham Lincoln issued his formal Emancipation Proclamation on January 1, 1863, giving freedom to slaves living in states in rebellion. This had little immediate effect in the South, but the impact overseas was decisive. No European state now could side openly with the Confederacy and thereby endorse human bondage.

In other ways the outlook was extremely bleak for the South. In the West, the Federals had wrested control of Kentucky and Tennessee and had seized all but a narrow stretch of the Mississippi River.

The Army of Northern Virginia also was much weaker than it had been only seven months previously, when Lee took command. True, in that time Lee's army had inflicted far more casualties than it had sustained: the North's casualties amounted to 70,725, the South's 48,171.[2] But his army had given up 4,077 prisoners while it had captured 29,370 Federal soldiers. Since prisoners were routinely exchanged, this meant that the *actual* Confederate losses, killed and wounded, were worse than actual Federal losses—44,094 to 41,345. With less than a third of the white population of the North, the South could not long endure Lee's method of waging war.[3]

Early in 1863 Lee also began to lose troops to peripheral dangers that absorbed the civilian leaders but had little significance in pursuing the war. The Federal garrison commander at New Bern, North Carolina, had raided westward in December, temporarily breaking the Weldon and Wilmington Railroad at Goldsboro, and arousing fear that the Federals were planning to close this vital Confederate supply line. On January 3, in response to demands from Richmond, Lee sent down Robert Ransom Jr.'s small division to block further attacks, and on January 14 he detailed native son D. H. Hill to direct military operations there.

In February the Federal 9th Corps, under William Farrar Smith, moved by sea to Fort Monroe, arousing alarm that the Federals would march up the James River Valley toward the capital or advance into the Carolinas. Lee sent George Pickett's and John Bell Hood's divisions, both from Longstreet's 1st Corps, south to meet Smith's force, and on February 17 he sent Longstreet himself. This left Lee with about 60,000 men and 170 guns to meet the Federal army of 138,000 men and 428 guns.

Although the 9th Corps soon departed for the West, General Longstreet became excited about Secretary of War James A. Seddon's suggestion to attack Suffolk, a few miles west of Norfolk. Though the 20,000-man Union garrison there interfered somewhat with Confederate commissary officers in securing hams and bacon from the rich hog-growing country west of Suffolk, it presented no strategic danger. Even if Longstreet succeeded in capturing Suffolk, it would have no effect on the Union Army of the Potomac, which was the real force targeting the Army of Northern Virginia.

Lee understood well the need to concentrate the Confederate army against its greatest danger. He wrote Gustavus W. Smith, commanding the Richmond defenses: "Partial encroachments of the enemy we must expect, but they can always be recovered, and any defeat of their large army will reinstate everything."[4] But President Davis, seconded by Seddon and Longstreet, ignored his pleas not to send troops to the south. As he was wont to do, Lee bowed to the civilian leadership—even though this leadership often was militarily inept and succumbed to pressures from politicians who, irrespective of the South's strategic needs, didn't want their territories occupied by Union troops.[5]

Longstreet demonstrated an unseemly willingness to sacrifice strategic considerations to his own ambition. As it turned out, he failed to capture Suffolk, yet by pursuing this plan he held more than 20,000 Confederate troops out of the titanic confrontation about to take place along the Rappahannock River. If present, these troops might have made a great difference in what the Confederates accomplished.

*   *   *

UNION GENERAL AMBROSE BURNSIDE recognized that his reputation in the army and in Washington had plummeted after his disastrous performance at Fredericksburg. Desperately trying to prevent Lincoln from removing him, he undertook to cross the Rappahannock over the narrow fords to the west of town on January 19, 1863. But violent rain fell on January 20 and continued for days, and the Federal advance stalled in deep mud.

The "mud march" was the last straw. Lincoln dismissed Burnside on January 25 and replaced him with "Fighting Joe" Hooker, forty-nine years old, a West Pointer of such monumental political ineptitude that not only had he criticized Burnside so openly that Burnside wanted him dismissed from the army, but he had denounced Lincoln as incompetent and suggested that the Union needed a dictator.[6]

Neither Lincoln nor anyone in his cabinet was impressed with Hooker's military abilities. His selection came almost wholly because he was free from political aspirations and was not a possible rival for Lincoln for the presidency in 1864.[7] Lincoln wrote him on January 26 that "only generals who gain successes can set up dictators. What I now ask of you is military success and I will risk the dictatorship."[8]

Thus Lincoln, concerned less with military competence than presidential politics, once more entrusted his principal army to an incapable leader. G. F. R. Henderson saw the error clearly. "The government which commits the charge of its armed forces to any other than the ablest and most experienced soldier the country can produce is but laying the foundation of national disaster," he wrote. "Had the importance of careful selection for the higher commands been understood in the North as it was understood in the South, Lee and Jackson would have been opposed by foes more formidable than Pope and Burnside, or Banks and Frémont." The Federal government, he wrote, considered that any ordinary general should be able to win victories.[9]

Lincoln turned to crises at home. The transformation of the war into a crusade to free the slaves virtually stopped volunteering for the army Lincoln responded by pushing through Congress a conscription act, which he signed into law on March 3, 1863. The administration applied the law to suppress dissent as well. Whenever army provost marshals encountered opposition, they jailed the disaffected person and denied him trial. This caused immense opposition and chilled Democrats who had previously supported the war effort. Leadership of the Democratic Party fell into the hands of people opposed to Lincoln, especially "Peace Democrats" such as Fernando Wood, a former mayor of New York City, and "Copperheads," especially in the Midwest, who were weary of the bloodshed and ready to end the war by compromise.

❋   ❋   ❋

THE FIRST CLASH in the 1863 campaign was a cavalry engagement on March 17 at Kelly's Ford, about twenty-three miles upstream from Fredericksburg. The Federal cavalry had improved greatly over the winter, now numbered 11,500 troopers, and did well against Jeb Stuart's 4,400 veterans. The gallant horse-artilleryman John Pelham died from a shell burst. But the Federal cavalry withdrew after the clash. The skirmish told Lee that Hooker was looking westward to make at least part of his advance.

Upstream the Rappahannock was narrower, the roads firmer, and the fords more accessible. Lee decided that the turning movement would be centered on United States Ford, about eleven miles northwest of Fredericksburg. He picked this ford because it gave access to Chancellorsville, three miles south, a crossroads where four roads came together, including the Orange Plank Road and Orange Turnpike (present-day Virginia Route 3).[10]

If Hooker got on the Plank Road and turnpike, therefore, he might press on Lee's flank and drive him out of Fredericksburg. Hooker's original plan, however, was to send the bulk of the Union army over the Rappahannock downstream from Fredericksburg, while crossing or pretending to cross with minor forces at United States Ford. He hoped to cut the Richmond, Fredericksburg and Potomac Railroad (RF&P), Lee's major supply line, and also to keep Lee away from the Union supply base at Aquia Creek, which Hooker believed, incorrectly, that Lee planned to strike.

Hooker also decided to send most of his cavalry southwest to seize Gordonsville and other points on the Virginia Central Railroad, then turn east and block the retreat of Lee's army, while cutting the RF&P near Ashland, just north of Richmond. He believed this action would force Lee to evacuate Fredericksburg and flee toward Richmond or Gordonsville. Since the inefficient Confederate supply system was able to keep only a few days' rations with the army, he reasoned, Lee would be obligated to fall back on his depots.[11] Hooker believed he could destroy Lee's retreating army by attacking it from the north while Union cavalry blocked its movement southward.

Hooker's idea of cavalry stopping a field army was strategic nonsense. The war had shown that horsemen could hold a blocking position only when dismounted, thereby sacrificing their mobility. The Confederates could reopen the railroads by sending infantry to oust any cavalry raiding force.[12] Besides, Lee had no history of retreating, and the Rebels were quite familiar with empty stomachs. If the Confederate army did not run, a cavalry strike would be a blow in the air. Worse, much of the Federal cavalry strength would be unavailable to serve in a pitched battle as Hooker's eyes and ears.

Hooker had expected his cavalry to move two weeks before the infantry advanced. But rains set in on April 15, making the Rappahannock unfordable. In response, Hooker abandoned his plan to strike around

Lee's eastern flank and substituted a major move around his western flank. Confederate Porter Alexander later called this substitution "decidedly the best strategy conceived in any of the campaigns set on foot against us."[13]

Hooker directed that two Union corps (the 1st and 6th), 40,000 men under Major General John Sedgwick, were to cross the river at Fredericksburg near where William B. Franklin's divisions had crossed in December, and hold in place the main Confederate army on the heights south of town. Meanwhile, three additional corps (the 5th, 11th, and 12th), 42,000 men under Major General Henry W. Slocum, were to march upstream and cross not at the obvious United States Ford, where Confederate outposts were on guard, but all the way up to Kelly's Ford, twelve miles above the junction of the Rappahannock and Rapidan rivers. From there the Union force would turn back down on the right or southern bank of the Rappahannock—a movement Lee would not expect that far upstream, since it would require fording both the Rappahannock and the Rapidan. Without opposition from Lee, the Federals could seize both United States Ford and Banks Ford, four miles west of Fredericksburg. Capturing these two fords would put the two wings of the Union army back into contact while at the same time unhinging the Rebel line at Fredericksburg.

Hooker intended to confront Lee in the broad, rolling, cultivated fields opposite Banks Ford. Here the Federal army could be fully deployed and here, in open countryside, it could mount its vastly superior artillery. In this situation, Hooker believed the Union army would be unbeatable and Lee would have to fight under impossible conditions, or retreat.

On April 27, the three corps under Slocum marched in great secrecy toward Kelly's Ford. The next day the 17,000-man 2nd Corps under Darius N. Couch—less John Gibbon's division, which was left at Falmouth to deceive the Confederates—marched to the north sides of United States and Banks fords. Couch's job was to convince Lee that he was preparing to force crossings at these two points. All of the Union army's reserve artillery moved up opposite Banks Ford as well, intending to cross as soon as Slocum's men secured the south bank.

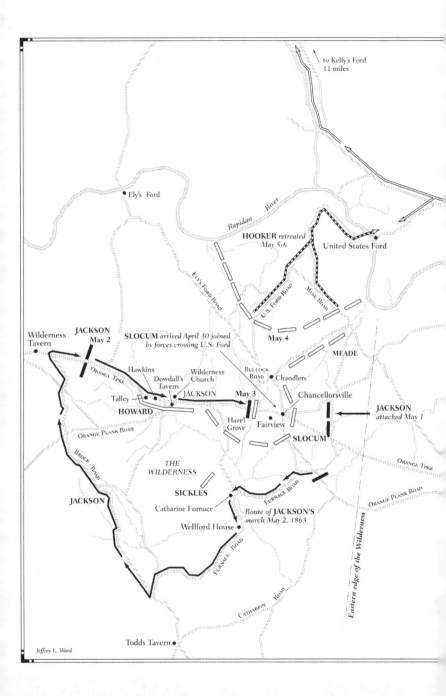

to Kelly's Ford
11 miles

Ely's Ford

Rapidan River

HOOKER *retreated*
May 5–6

United States Ford

ELY'S FORD ROAD

U.S. FORD ROAD

MINE ROAD

May 4

MEADE

Wilderness
Tavern

JACKSON
May 2

ORANGE TPKE

SLOCUM *arrived April 30 joined
by forces crossing U.S. Ford*

Hawkins

Dowdall's
Tavern

Wilderness
Church

JACKSON

BULLOCK
ROAD

Chandlers

May 3

Chancellorsville

JACKSON
attacked May 1

Talley

HOWARD

ORANGE PLANK ROAD

Hazel
Grove

Fairview

SLOCUM

ORANGE TPKE

BROCK ROAD

THE
WILDERNESS

SICKLES

Catharine Furnace

FURNACE ROAD

Route of JACKSON'S
march May 2, 1863

ORANGE PLANK ROAD

JACKSON

Wellford House

FURNACE ROAD

Eastern edge of the Wilderness

CATHARPIN ROAD

Todds Tavern

Jeffrey L. Ward

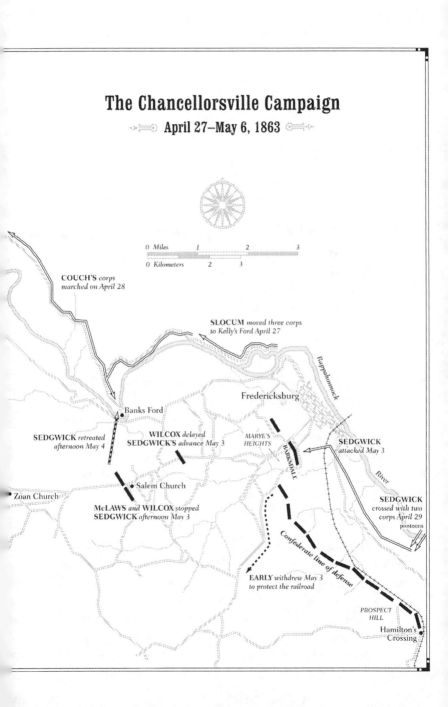

The Chancellorsville Campaign
April 27–May 6, 1863

0 Miles 1 2 3
0 Kilometers 2 3

COUCH'S *corps marched on April 28*

SLOCUM *moved three corps to Kelly's Ford April 27*

Fredericksburg

Rappahannock

Banks Ford

SEDGWICK *retreated afternoon May 4*

WILCOX *delayed SEDGWICK'S advance May 3*

MARYE'S HEIGHTS

BARKSDALE

SEDGWICK *attacked May 3*

River

Zoan Church

Salem Church

McLAWS *and* **WILCOX** *stopped SEDGWICK afternoon May 3*

SEDGWICK *crossed with two corps April 29* pontoons

Confederate line of defense

EARLY *withdrew May 3 to protect the railroad*

PROSPECT HILL

Hamilton's Crossing

The same day, April 28, Rebel outposts reported to Lee the movement on the western flank, but the Confederate commander believed the force was moving on the Shenandoah Valley.

Early on April 29, Sedgwick crossed the Rappahannock with his two corps about three miles below Fredericksburg. His troops dug in at the riverbank opposite Stonewall Jackson's corps. They made no attempt to storm either Prospect Hill or Marye's Heights. Hooker's last corps, the 3rd, 18,700 men under Major General Daniel E. Sickles, remained at Stafford Heights, ready to reinforce either Slocum or Sedgwick, as required.

Meanwhile, 10,000 Union cavalrymen under George Stoneman crossed the river upstream and struck out on a wide-ranging raid into Virginia. Hooker left only 1,300 cavalrymen under Alfred Pleasonton to scout ahead of Slocum's infantry. Jeb Stuart at first could not figure out the Union plan, guessed that Slocum's force was only one-third its actual size, and thought it was heading south toward Gordonsville. He posted his patrols accordingly.

Thus, by the morning of April 29, all three Union corps were across Kelly's Ford and had turned back east without being detected by Stuart's cavalry. The Federals waded across two fords on the Rapidan, Germanna (on present-day Virginia Route 3) and Ely's (on present-day Virginia Route 610), about ten miles east of Kelly's.

Couriers alerted Lee on the afternoon of April 29 that the Federals had passed these fords. Lee now realized that Stuart was wrong and that a Union force was descending on his left flank. He still didn't know its size. Even so, he at once sent Richard H. Anderson's division to the Chancellorsville crossroads.

Stuart, realizing he was out of place, moved with Fitzhugh Lee's 3,300-man brigade eastward to assist Lee and detailed W. H. F. "Rooney" Lee's brigade of only 1,000 troopers to deal with Stoneman's 10,000 Union cavalry in the Confederate rear. Stuart thus concentrated the bulk of his horsemen where they were most needed—at the front of both armies. This gave Lee quick and accurate information, while Pleasonton, who could not penetrate Stuart's shields, was unable to inform Hooker of the

Confederate movements. Lee ignored the breaks of the railway lines in his rear, and Stoneman's move degenerated into a giant but useless raid.

By the morning of April 30, Stuart had captured prisoners from the three Union corps, and Lee now knew the magnitude of the force opposing him. It was two-thirds the size of his whole army. It was obvious that Anderson could not stand alone. Lee ordered him to select a strong position and dig in. Anderson retreated back four and a half miles east of Chancellorsville to Zoan Church and began to build entrenchments.[14]

Soon afterward, Slocum's advance party arrived at Chancellorsville. The only structure at the crossroads was the large two-story brick Chancellor house, with pillars and an extensive clearing around it. Slocum's orders from Hooker were to continue eastward to seize Banks Ford, about six miles northeast of Chancellorsville.

The remainder of the three Union corps closed up on Chancellorsville that day. They had marched forty-six miles and had achieved a spectacular surprise. Heavy reinforcements were coming by way of United States Ford, which Couch's soldiers had occupied. Two divisions of Couch's 2nd Corps and all of Sickles's 3rd Corps would arrive on the morning of Friday, May 1, giving Hooker 70,000 men and 208 guns on Lee's flank.

But Hooker would not achieve complete success until he had secured Banks Ford, which would reconnect the two wings and also get the 70,000 men at Chancellorsville out of a labyrinthine, dangerous region known as the Wilderness.

The eastern edge of the Wilderness lay a couple of miles east of Chancellorsville. The original forest stretched fourteen miles along the Rappahannock and eight to ten miles to the south. It had been cleared over the previous half century to supply charcoal for iron furnaces in the area. A dense second growth of low-branched hardwood, pine, and cedars, as well as briars and underbrush, had created a tangled, difficult landscape with few roads, poor visibility, and only a few small open places. In this Wilderness, Hooker's greatest asset, his artillery, was useless.

On April 30, Lee faced two imminent dangers: a huge force on his left flank and another huge force on the south bank of the Rappahannock. If Hooker, on the west, advanced much farther, he would possess Banks

Ford and be free of the Wilderness, and could strike a perhaps mortal blow against Lee's left flank.

The principal reason why Hooker had moved against Lee's left flank was to strike at a weakly held point. But it would remain weakly held and vulnerable only if Sedgwick sent in a powerful frontal assault against the main Confederate line at Fredericksburg. If Sedgwick assaulted, Jackson would be compelled to use nearly all his troops to defend Marye's Heights and Prospect Hill.

To the astonishment of Lee and Jackson, Sedgwick did not attack. The reason was that Hooker had given Sedgwick the option of deciding whether or not he would advance. He did not insist on an all-out attack. This was a fatal error. Sedgwick decided not to move, and only hunkered down along the banks of the river.

Seeing that Sedgwick remained idle, Lee ordered Lafayette McLaws's division to march at once to aid Anderson at Zoan Church, leaving only William Barksdale's Mississippi brigade to defend Marye's Heights. He also ordered Jackson to march three divisions of his corps at daylight the next day, May 1, to Zoan Church, take charge of the western flank, and "repulse the enemy." Jackson left his fourth division, under Jubal Early, along with William N. Pendleton's reserve artillery, to watch Sedgwick. Thus at nightfall on April 30, Lee had turned his back on Sedgwick's huge force of 40,000 men and concentrated 47,000 men and 114 guns on the west to confront Hooker. He left only about 10,000 men on the Fredericksburg heights under Early.

Hooker had completely misconstrued what Lee would do. His order of the day on April 30 read: "Our enemy must either ingloriously fly or come out from behind his defenses and give us battle on our own ground, where certain destruction awaits him."[15] But far from ingloriously flying, Lee was turning every possible man and gun he had against the challenge at Chancellorsville.

Hooker had planned only a modest advance on May 1, to get out of the Wilderness and seize Banks Ford. Hooker started this move in the morning, sending separate columns eastward on the Orange Plank Road to the south, the turnpike in the center, and the River Road to Banks Ford to the north.

But Stonewall Jackson had already reached Zoan Church with his corps at 8:00 A.M. He was well aware of the character of the Wilderness. He also knew the Confederates would be unable to counter Federal cannons if they got into the open east of the Wilderness. Accordingly, the moment he arrived, he told Anderson and McLaws to stop building entrenchments and, with his corps, to form up in order of battle to advance westward into the Wilderness.

Jackson saw that if he could push Hooker back into the Wilderness, the Federal artillery would be badly crippled. Though the woods would provide excellent defensive positions, bottling up the Union guns would more than even the odds. More important, by pushing Hooker back into the Wilderness, he would prevent the Union army from reaching open country, where its vastly greater power might overwhelm the pitifully small army Lee now had under his command. In a stroke, Jackson turned a desperate situation that threatened the destruction of the Army of Northern Virginia into an opportunity for victory.

Hooker was stunned by Jackson's move. Instead of using his immensely superior force to challenge Jackson's advance and drive him back, Hooker withdrew to Chancellorsville and built a defensive line in an arc of crude but strong earthworks and logs just east and south of the crossroads.[16]

It was an unbelievable retreat. His generals were deeply angered. When General Couch reported to Hooker on the night of May 1, Hooker said, "I have got Lee just where I want him. He must fight me on my own ground." Couch later wrote: "To hear from his own lips that the advantages gained by the successful marches of his lieutenants were to culminate in fighting a defensive battle in that nest of thickets was too much, and I retired from his presence with the belief that my commanding general was a whipped man."[17]

Another indication of Hooker's distraction was that on the night of May 1 he detached the 16,000-man 1st Corps under John F. Reynolds from Sedgwick's command and ordered it to march to Chancellorsville. This left Sedgwick with 24,000 men in his 6th Corps, and indicated that Hooker had abandoned any thought of Sedgwick holding the bulk of Lee's army on the heights below Fredericksburg.

Porter Alexander remarked on "the perfect collapse of the moral courage of Hooker, as commander in chief, as soon as he found himself in the actual presence of Lee and Jackson."[18]

※   ※   ※

GENERAL LEE ARRIVED in the afternoon of May 1, and he rode out to the north to determine whether there was any chance of turning the Federal line there and cutting the Union army off from United States Ford. But Lee found too many Federal troops and too little space to maneuver around the northern flank.

Around 7:30 P.M. Lee met with Jackson. The Army of Northern Virginia had only two options—to attack Hooker's positions at Chancellorsville frontally or to move around the southern Union flank. Lee sent off two engineers to study Hooker's defensive positions. They reported back that they were entirely too strong to be carried by frontal assault.

This left only a flanking movement. As Lee and Jackson were discussing the matter, Jeb Stuart rode up and announced that Fitzhugh Lee had discovered the Federal right stretched out along the Orange Turnpike west of Chancellorsville, facing south. Moreover, this line, he said, was "floating in the air," meaning that it rested on no secure defensive position on its western end, and the corps commander there, Oliver O. Howard, had established no defenses facing west.

Here was an incredible bonanza. If Confederate troops could swing all the way around Hooker's southern flank and emerge on the turnpike facing east, the Confederates could drive straight down the road toward Chancellorsville and roll up the entire western flank of the Union army.

The prospect presented great danger, however. The Confederate army was only half the size of the Union army, and it had already been divided in part because Lee had to leave 10,000 men to watch Sedgwick at Fredericksburg. To send a Rebel force around Hooker's southern flank would require dividing the army once more. Either segment, left alone, would be too small to fight a pitched battle if Hooker struck with most of his force.

Even so, both Lee and Jackson recognized that a flanking movement presented the only means to drive Hooker back across the Rappahannock. Lee ordered the operation and appointed Jackson to carry it out, with Stuart to shield the march with his cavalry. Jackson rose, smiling, touched his cap, and said, "My troops will move at 4 o'clock."

Although Lee had approved the turning movement, nothing else had been decided. The route of the march, the exact objective, and the number of troops to be used were still to be settled.

Lee's original idea was for a simple flanking movement to dislodge Hooker from his defensive positions around Chancellorsville and force him to retreat back across the river. But Jackson saw a way to destroy the Union army. If Confederate troops could get between United States Ford and Hooker's forces around Chancellorsville, they could cut off his only means of retreat—and the Union army, caught between Jackson barring passage on one side and Lee pressing on the other, would be compelled to surrender.

The ideal movement, therefore, would be to press from Hooker's exposed western flank straight northeast to United States Ford. But this route was barred by the absence of any direct roads through the Wilderness. The only feasible path was to drive directly east down the Orange Turnpike to Chancellorsville. Just west of this crossroads, the Bullock Road led northeast less than a mile to Chandlers or White House, on the main road leading north to United States Ford. If the Confederates could get to Chandlers, they would block Hooker's main retreat route. And from Chandlers they could push northeast for two more miles and block Mine Road, the only other route leading to the ford.

Jackson himself specifically stated that this was his aim. This was the instance in which we have a record of his articulating his theory of warfare. After the battle, Jackson's medical officer, Hunter McGuire, reported that "he told me that he intended, after breaking into Hooker's rear, to take and fortify a suitable position, cutting him off from the river and so hold him until, between himself and General Lee, the great Federal host should be broken to pieces. He had no fear. It was then that I heard him say, 'We sometimes fail to drive them from position, they always fail to drive us.'"[19]

Early on the morning of May 2, Jackson awoke from a short sleep and questioned his chaplain, Tucker Lacy, whose family owned land in the region, about a route to follow. Lacy remembered that Charles C. Well-ford, owner of an iron furnace known as Catharine Furnace, lived a couple of miles southwest. Jackson at once called his mapmaker, Jede-diah Hotchkiss, to seek out Wellford. Hotchkiss and Lacy woke Wellford, who pointed out a covered route and appointed his young son, Charles, as guide.

By the time Hotchkiss returned, Jackson was conferring again with Lee. Hotchkiss traced the route recommended by Wellford. There was a moment of silence. Lee then said: "General Jackson, what do you pro-pose to do?" Jackson replied: "Go around there," pointing to the line Hotchkiss had shown. Lee: "What do you propose to make this move-ment with?" Jackson, without hesitation: "With my whole corps." Lee replied: "What will you leave me?" Jackson: "The divisions of Anderson and McLaws."

This was the moment of decision in the Civil War. Jackson was pro-posing anything but a normal turning movement with a small portion of the Confederate army. He was proposing a massive strike to achieve a resolution of the war. He was proposing that the bulk of the remaining forces Lee possessed, about 25,000 men, be used. Because of the size of this force, it was obvious that Jackson intended to block Hooker's pas-sage to the river, not merely drive him back across the river. The Army of Northern Virginia—the entire Southern cause—was being risked. But Lee saw that Jackson was utterly confident. For Jackson it was not a risk at all. To him it was a dead certainty. This was the strategy Jackson had been preaching ever since he tried to get Lee to swing around Pope on the Rapidan nine months before. He was confident that if he could block the Union army's line of retreat, he could hold, and "the great Federal host" would be broken to pieces.

Lee had rejected Jackson's earlier proposals for a massive strike on the enemy's flank. But here was an opportunity that might never come again. Hooker had placed himself in a perilous position, with only one river crossing. Lee recognized that Jackson had seen a chance to trans-form the situation. He hesitated only briefly, then said calmly, "Well, go on."[20]

Jackson's corps began to move forward around 7:00 A.M. As the head of the column swung southwest toward Catharine Furnace, Stonewall Jackson rode a short distance behind with his staff. Lee stood by the road to say goodbye. Jackson drew rein and they talked briefly. Jackson pointed ahead, Lee nodded, and Jackson rode on. One of the most spectacular marches in the history of warfare had begun.

✤   ✤   ✤

FOR MOST OF Jackson's soldiers, the actual march around Hooker was uneventful, a long, hard trek of about twelve miles along narrow dirt roads cut through the forest but with little dust because of recent rains. Stuart's horsemen shielded them from any disturbance from Federal patrols. The men were eager and expectant. Jackson was leading them off on another of his mysterious marches, and they were delighted with the prospect.

The forces left with Lee, about 18,000 men, started demonstrations to make the Federals believe an attack was coming from the east. Nevertheless, the Federals discovered Jackson's march soon after it started. Brigadier General David B. Birney, commanding a division in Daniel Sickles's 3rd Corps, had forces on a cleared elevation known as Hazel Grove, a little over a mile southwest of Chancellorsville. About 8:00 A.M. some of his scouts reported passage of a long column heading southwest on the Furnace Road running past Hazel Grove.

Birney reported the news. Hooker decided that the Confederates were retreating. The direction indicated Gordonsville as the destination. Although he sent a warning to Henry W. Slocum, commanding the 12th Corps, and Oliver O. Howard, commanding the 13,000-man 11th Corps, about a possible attack on their flanks, he did not regard the threat as great, and neither Slocum nor Howard gave it much attention.

Birney ordered a Federal battery deployed at Hazel Grove to open fire on a spot on the road where the Rebel column was visible. This caused the diversion of Jackson's wagons to another, unexposed road farther east. Around noon Hooker authorized Sickles to advance on Catharine Furnace to attack Lee's "trains."

Birney's and Amiel W. Whipple's divisions moved forward, but Carnot Posey's Mississippi brigade in Richard Anderson's division, posted just east of the furnace, challenged the advance, while the 21st Georgia, detached from Jackson's column, defended the furnace. These Confederate forces allowed the remainder of Jackson's corps to pass beyond reach of the Federal probes.

After marching to the end of the Furnace Road, three and a half miles southwest of the Orange Plank Road, Jackson's column turned back northwest on the Brock Road (present-day Virginia Route 613). About 1:00 P.M. some 2nd Virginia Cavalry horsemen went east up the Plank Road and saw the right of the 11th Corps's line. Jackson and Fitz Lee rode up the Plank Road to an elevated point a mile southwest of Dowdall's Tavern, near the junction with the turnpike. Jackson saw that the Union entrenchments lay a few hundred yards away, facing south.

Jackson had planned to turn up the Plank Road to attack, thinking that the end of Howard's positions lay farther east. Now he continued north on the Brock Road for another mile and a half to the turnpike, then turned east, certain he then would be beyond the farthest western Federal element.

As Jackson was completing his march around Hooker, General Sickles decided he could break the Confederate column, which he and Hooker still believed was retreating toward Gordonsville. He surrounded the 21st Georgia at Catharine Furnace, capturing most of the members, but, fearing Rebel rear guards would be alert, he called for reinforcements. Per Hooker's order, Howard sent Francis C. Barlow's 1,500-man brigade to the furnace.

Howard's corps was thus reduced to 11,500 men. It was spread out for nearly two miles along the Orange Turnpike. Most emplacements faced south, not west, though the most westerly portion was facing west—two regiments of Leopold von Gilsa's brigade, from Charles Devens's division. They were behind a weak abatis of felled small trees and bushes. This emplacement was just north of the turnpike and a little more than a mile west of Dowdall's Tavern. With these two regiments were two cannons that were deployed to fire on the road, not to the west. Another regiment, the 75th Ohio, was some distance east of von Gilsa, in reserve. The remainder of Devens's division was on either side of Talley's house and farm, three-quarters of a mile west of Dowdall's Tavern. Devens's men faced south along the pike.

About a quarter of a mile northeast of the Talley farmstead was the Hawkins farm, on cleared high ground. Here was Carl Schurz's two-brigade division.[21] Most of this division also faced south, but two regiments of Alexander Schimmelpfennig's brigade faced west. It also was shielded by a light abatis.

The final brigade, commanded by Adolphus Buschbeck, of Adolph von Steinwehr's division, was facing south at Dowdall's Tavern, where a considerable area was clear of trees. Here a shallow shelter trench or ditch ran several hundred yards, facing west just north of the road, with several cannons behind it. The trench, however, was unoccupied.

Between Howard's 11th Corps and Slocum's 12th Corps around Chancellorsville was a mile of unoccupied ground, previously held by Barlow's brigade, now marching to aid Sickles. Below Catharine Furnace were two divisions of Sickles's corps, Barlow's brigade and one regiment of Alfred Pleasonton's cavalry, perhaps 16,000 men, who thought they were pursuing Lee's fleeing army.

In summary, two weak lines of Howard's corps faced west, but most of his men and guns were facing south. Sickles's corps was so far south as to be beyond the fight. The remainder of Hooker's army faced largely east or south, defending against Lee's two divisions, and was in no position to block Chancellorsville from the rear. Only one division, Hiram G. Berry's of Sickles's corps, was in the vicinity of the crossroads and might be called on in an emergency.

Jackson's plan was to move eastward along the turnpike, roll up Howard's 11th Corps, and drive into the rear of the corps of Couch, Sickles, and Slocum—primarily to get to the crossroads of Chandlers, just north of Chancellorsville.

Jackson deployed his men, making as little noise as possible. In the first line went Robert E. Rodes's division. In the second line, 200 yards back, was Raleigh E. Colston's division. Behind, partly in column, came A. P. Hill's division.

Since Howard's corps was on the turnpike, the main Confederate thrust was going to be made by three brigades of Rodes's division near the road—George Doles's Georgia brigade, lined up just south of the road; Edward A. O'Neal's Alabama brigade, just north; and Alfred Iverson's North Carolina brigade, to O'Neal's left. On the right or south, in the first line was Alfred H. Colquitt's Georgia brigade. Directly behind was S. Dodson Ramseur's North Carolina brigade. A short distance to the south was E. P. Paxton's Stonewall Brigade of Colston's division, positioned to march straight up the Plank Road to Dowdall's Tavern as the assault lines moved out. Paxton's job was to clear out any Union detachments that might be south of the main line along the turnpike.

Neither Colquitt's, Ramseur's, nor Paxton's brigade was likely to face significant Federal forces. Their advance was important, however, because their path would lead over both Hazel Grove and Fairview, an elevated, open area about two-thirds of a mile southwest of Chancellorsville. These two heights were the only places where many guns could be emplaced, and their capture would imperil Hooker's entire position. Also, seizing Hazel Grove would separate Sickles's large force from Hooker's main body, and probably lead to its surrender.

Jackson did not realize the significance of Hazel Grove and Fairview at the moment, but he ordered all of his troops to push resolutely ahead, to allow nothing to stop them, even disorder in their ranks. He directed that if any part of the first line needed help, it could call on aid from the second line without further instructions. Under no circumstances was there to be a pause.

About 5:15 P.M. all was ready, and Stonewall Jackson released his excited and enthusiastic soldiers. They descended like thunder on the Union army, which became aware of its danger only when deer, wild turkeys, and rabbits, stirred up by the Confederate lines, rushed in fright through their positions.

Doles's Georgians, about a mile forward from their starting point, encountered von Gilsa's soldiers preparing their evening meal. The Federals hastily formed a line of battle. Doles smashed straight into the position. The Federals stood three volleys but then fell apart, the men hurtling backward in complete disarray. Von Gilsa's regiments facing south, hit by volleys from front, flank, and rear, disintegrated without firing a shot.

A few Union soldiers rallied around the 75th Ohio, but it, too, turned and fled.

The great bulk of Devens's division, facing south, abandoned their positions and ran headlong toward Chancellorsville. General Howard, watching the disaster unfold from the elevation around Dowdall's Tavern, noted that "more quickly than it could be told, with all the fury of the wildest hailstorm, everything, every sort of organization that lay in the path of the mad current of panic-stricken men had to give way and be broken into fragments." Howard's aide was struck dead by a shot, and Howard's horse sprang up on its hind legs and fell over, throwing the general to the ground.[22]

As Rodes's three Rebel brigades pressed eagerly toward Dowdall's Tavern, Colquitt on the south advanced only a few hundred yards, then halted, in direct disobedience of orders. He had gotten a report that Federals were on his southern flank. Paxton's brigade was positioned to deal with any stray Union soldiers in the area, but Colquitt's halt forced Ramseur's and Paxton's brigades to halt as well. Ramseur, exasperated, assured Colquitt that he would take care of any Federals that appeared. At last Colquitt moved, but three brigades, 5,000 men, were now so far behind Jackson's advance that they couldn't catch up. Colquitt's disobedience prevented the Confederates from seizing Hazel Grove and Fairview and from severing Sickles's large force from the main army. After this battle, Lee would ship the disgraced Colquitt south, swapping his brigade for a North Carolina force.

The last organized Federal force in front of Jackson was Buschbeck's, at Dowdall's Tavern. Buschbeck's men had moved into the shallow trench and were facing westward, but they were nervous and tentative as they watched the huge Rebel force descending on them. Jackson assailed Buschbeck's line along its whole front, while he rolled additional troops around each flank. A sheet of fire struck some Federals in the trench. As they went down, men on either side vacated the trench and ran away, many throwing away their arms and joining the chaotic stream of men, horses, cannons, and wagons rushing to the rear.

But around 7:15 P.M., about a mile and a half west of Chancellorsville, Rodes, an orderly man not gifted with the vision of his commander, called an abrupt halt to the advance of the Confederate line. The division, he decided, was too mixed up. Rodes's and Colston's lines were in fact getting intermixed, but the Rebels still advanced steadily. Rodes sent word to Jackson to send forward A. P. Hill's division, while he took his division back to Dowdall's Tavern to re-form.[23]

It is often the case in war that subordinate generals' foolish or thoughtless decisions vastly or even fatally damage a commander's brilliant vision. This is precisely what Robert Rodes and Alfred H. Colquitt did. Rodes's decision was particularly damaging, as it came at the most critical moment in the attack on Hooker. Since darkness was falling, it was imperative to reach and seize the Chandlers crossroads while it could be easily located. The unnecessary stop Rodes ordered gave the Federals an opportunity to take a breath and organize a defense. Moreover, it ended any chance of resuming the advance quickly, because it took until nightfall for A. P. Hill to bring up any troops.[24]

Hooker, at the Chancellor house at the crossroads, did not get news of Jackson's attack until 6:30 P.M., when Captain Harry Russell, one of his aides, heard violent turmoil and, turning his glass westward, yelled, "My God, here they come!" Russell believed the fleeing Federals were part of Sickles's corps. Only when Hooker and his aides rushed into the mass did they discover the truth. Hooker nearly panicked, sending word to Sickles to save his men if he could. If Colquitt had not stopped, Sickles could have been cut off from the main army.

Because of Rodes's halt, Hooker had an hour to stem to some degree the rout of his army. Hiram Berry's division was near, and Hooker ordered it to move west along the turnpike and challenge Jackson. Berry's men advanced resolutely through the panic-stricken mass of fleeing men and, around 8:00 P.M., started entrenching in the valley of a small stream about a half mile west of Chancellorsville, and just north of Fairview. At Fairview twenty Federal artillery pieces were being unlimbered and pointed westward.[25]

As Sickles's men rushed northward, Pleasonton organized a defense around some artillery at Hazel Grove. These guns held off a few Confederates from A. P. Hill's division who finally reached Hazel Grove in the darkness. In this fashion, Sickles's corps reconnected to Hooker's main force.

A. P. Hill was able to bring up only James Lane's North Carolina brigade immediately. It was 8:45 P.M. before the brigade was lined up on either side of the turnpike a mile west of Chancellorsville.

General Jackson arrived at the front. His intention was to send part of Hill's division northeast to seize Chandlers crossroads by way of Bullock Road, running directly to it from the advanced Confederate positions. Jackson still had time to cut off the Union army from the Rappahannock. The night was clear, with a full moon, giving sufficient light to move. Moreover, the Union soldiers were thoroughly demoralized and could have offered little resistance.

Jackson had not yet established a connection with Lee, which he meant to do by pushing Lane's brigade straight forward to and through Chancellorsville. When Lane asked for orders around 9:00 P.M., Jackson raised his arm in the direction of the enemy and exclaimed, "Push right ahead, Lane, right ahead."

Soon thereafter Hill arrived. "Press them," Jackson ordered him. "Cut them off from the United States Ford, Hill; press them."[26] Hill told Jackson that he was unfamiliar with the country leading to the ford, so Jackson ordered Captain J. Keith Boswell, well acquainted with the roads and paths of the Wilderness, to guide him.

Jackson went with Hill and his staff in advance of Lane's line to get further knowledge of the terrain in the direction of United States Ford. They heard Federal voices and axes cutting trees for abatis. Firing erupted in front of them. Jackson and his party hurriedly moved off the turnpike and turned into the little-used Mountain Road only a few yards north.

A short time previously, Union General Pleasonton, fearing that the Rebels were about to overrun Hazel Grove, had ordered the 8th Pennsylvania Cavalry to charge the enemy. The lead squadron collided with the front ranks of the Confederates on the turnpike. Numerous Union horsemen went down from Rebel volleys. The attack fell apart and the survivors fled, but riderless horses and horseless riders rushed about in the dark woods for some time.

Lane's soldiers were still alert from the cavalry charge when Jackson's party rode along the Mountain Road. The 18th North Carolina mistook the sounds for Federal cavalry. The order was given to fire. Jackson's party was not more than twenty paces away. Captain Boswell and an orderly fell dead. Three bullets struck Jackson. One penetrated his right palm; another went into the wrist of his left hand; a third splintered the bone of his left arm between the shoulder and the elbow, severing an artery. Jackson's horse bolted, but Jackson was barely able to hold it in check.

Jackson weakened rapidly from loss of blood, though Hill and his aides stopped some of the bleeding with a knotted handkerchief. After getting Jackson back to the medical station, Jackson's medical officer, Hunter McGuire, and other doctors amputated his left arm just below the shoulder.

Shortly after Jackson was hit, Union artillery fire wounded General Hill in the legs. He relinquished command to Rodes, and then to Jeb Stuart, now the only unwounded major general on the western flank. Stuart did not take over command until after midnight, however, because he was at Ely's Ford at the time.

The transfer of command was fatal to Jackson's plan to block Hooker's retreat to United States Ford. Stuart suspended operations until daylight. By then it was too late. Hooker already had George G. Meade's 5th Corps north of Chancellorsville, and John F. Reynolds's 1st Corps came up during the night. These two corps, 30,000 men, plus 25,000 Union soldiers Hooker got lined up west of Chancellorsville, were more than enough to overwhelm Stuart's entire force. Hooker did not, however, think of turning the tables on the Confederates. He ordered the 1st and 5th Corps to build a defensive line to protect United States Ford.

Lee realized that the time for Jackson's blocking United States Ford had passed. With only a cavalry major general and infantry brigadiers in command, the sooner he could reunite the two wings of the army, the better. He directed Stuart to press eastward, not to try to get athwart the roads to United States Ford. The effort to move eastward set in motion a bloody series of frontal assaults that succeeded at last, but cost both sides immense casualties.

Hooker now ordered the withdrawal of his entire force northward.

This withdrawal allowed the two wings of the Confederate army to reunite.

※　※　※

UNION GENERAL John Sedgwick finally moved from Fredericksburg on the morning of May 3. Sedgwick had 24,000 men facing Jubal Early's 3,500 men on Prospect Hill, southeast of Fredericksburg. Sedgwick could easily have driven through Hamilton's Crossing, turned the entire Confederate position, and struck straight for Lee's rear. Instead, he marched up the plain in front of the Confederate positions. There, joined by John Gibbon's 6,000-man division from Falmouth, he assaulted frontally the same Sunken Road below Marye's Heights that had ruined Ambrose Burnside's offensive in December 1862. As John Bigelow Jr., historian of the Chancellorsville campaign, wrote, "The result was the singular spectacle of a body of troops practically on the enemy's flank moving to the enemy's front in order to attack him."[27]

Unlike Burnside's assault, however, Sedgwick's offensive faced only a single Confederate brigade, William Barksdale's Mississippians. The first Union assaults failed, the Federals losing nearly 1,000 men in less than five minutes. Then Thomas M. Griffin, commanding the 18th Mississippi in the Sunken Road, foolishly accepted a Federal cease-fire to remove the wounded. This permitted Union officers to observe how few Rebels were defending the road. The Federals then attacked in heavy force and captured or destroyed nearly the whole Mississippi regiment.

Early withdrew down the Telegraph Road toward Richmond to protect the RF&P Railroad. Only Cadmus M. Wilcox's Alabama brigade of Anderson's division, guarding Banks Ford, was in position to delay Sedgwick's advance. But Sedgwick took so long to organize a strike west on the Plank Road and turnpike that Wilcox was able to form a strong defensive line six miles east of Chancellorsville at Salem Church. McLaws's division came up to help. Although Sedgwick had twice as many troops as faced him, he remained immobile on May 4, giving Lee the opportunity to organize a converging assault that drove Sedgwick across Banks Ford by early evening.[28]

On the night of May 5–6, Hooker took advantage of a huge storm to pull his entire force north of the Rappahannock. He marched his beaten army back to Falmouth, and Lee returned to the heights behind Fredericksburg.

※   ※   ※

BECAUSE OF THE wounding of Jackson, the South had lost its opportunity to destroy the Army of the Potomac. Now Jackson was dying. He had been moved back to Guiney's Station (now Guinea) on the RF&P to recuperate. But he contracted pneumonia. Jackson lingered until Sunday, May 10, 1863. In his last hours his mind wandered back to battle, and he called out: "Order A. P. Hill to prepare for action! Pass the infantry to the front! Tell Major Hawks—" He became silent for a while, then said quietly and clearly, "Let us cross over the river and rest under the shade of the trees." Stonewall Jackson was dead.[29]

The South had achieved its most spectacular victory in the war. Yet it paid too high a price for Chancellorsville, for it lost Stonewall Jackson. And it had no other commander with anything like the capacity for victory that Jackson had possessed.

Union General Oliver O. Howard described the South's loss best: "Providentially for us, it was the last battle that he waged against the American Union. For, in bold planning, in energy of execution . . . in indefatigable activity and moral ascendancy, Jackson stood head and shoulders above his confrères, and after his death General Lee could not replace him."[30]

Jackson had recognized the one necessity of successful warfare: victory requires avoiding the enemy's strength and striking where the enemy is not. Jackson had also found a way to avoid making frontal attacks against the massed power of the much larger Union army, by getting his enemy to attack Confederate positions. These two insights were the essence of Jackson's intellectual breakthrough. But Lee, despite his acceptance of Jackson's plan at Chancellorsville, had not absorbed the lesson. And this sealed the fate of the Confederacy.

Gettysburg

HE WAR HAD entered a decisive phase. The South had achieved its greatest victory at Chancellorsville. It was incomplete, in that it did not destroy the Union army, but nevertheless convincing. Ominous signs pointed to impending Southern weaknesses, however. Stonewall Jackson was dead. Ulysses S. Grant was threatening Vicksburg, the Confederacy's last strongpoint on the Mississippi and its only rail connection with the trans-Mississippi states. Days after Lee and Jackson had won at Chancellorsville, Grant had evicted a Southern garrison at Grand Gulf on the Mississippi and started marching toward Vicksburg, twenty-five miles north.

Grant's move set off an immense fright in Richmond. Secretary of War James A. Seddon diverted troops to reinforce John Clifford Pemberton, the Confederate commander at Vicksburg, and ordered General Joseph E. Johnston to take charge personally.

Although Confederate civilian leaders were distraught at the thought of the trans-Mississippi states being cut off from the rest of the South, the threat did not disturb the senior Confederate military leaders. Goods coming from the region were so few that their loss would not cripple the Confederacy. General James Longstreet felt the war would be little affected by the capture of Vicksburg. Lee saw the city only as a means of interdicting Federal traffic on the river. Johnston regarded middle Tennessee and the threat to Chattanooga as much more important.

But President Jefferson Davis tried to get Lee to send some of his troops to defend Vicksburg. Despite Lee's habit of bowing to the requests of the president, he resisted this proposal fiercely and said the Army of Northern Virginia instead should be reinforced, pointing out specifically that thousands of Confederate troops were standing idle, guarding southern coasts that were never going to be invaded. These, he said, should be brought north to reinforce his army. The only result, however, was to convince Davis to send back the divisions of George Pickett and John Bell Hood, which had been detached during the winter.

When Grant moved to Jackson, Mississippi, forty-five miles east of Vicksburg, he drove out a Confederate force under Joe Johnston, cut the railroad to Vicksburg, and isolated the city. General Pemberton, instead of abandoning the city and joining his troops with those of Johnston, withdrew to Vicksburg, which Grant immediately besieged. Although Johnston had 31,000 troops, he made no active effort to break the siege, despite appeals from Richmond. As the distinguished turn-of-the-century military historian Theodore Ayrault Dodge wrote, "Johnston, with all his ability, was never distinguished as a fighter. His tendency was dilatory rather than active."[1] Pemberton's foolish decision therefore spelled not only the loss of the city but also the surrender of its entire garrison.

✸   ✸   ✸

LEE DECIDED TO invade the North once more. His aim was to achieve a decision in the war. But just as he'd done in his 1862 incursion into Maryland, Lee disguised his intentions with Davis, who still believed the South could outlast the North. Lee knew Davis's approach was a recipe for defeat, and he had come to agree with Stonewall Jackson—that the only way to induce Lincoln's government to quit would be to strike at the North, and ultimately defeat a Union army on Northern soil.[2]

Lee met with Davis's cabinet in Richmond on May 16, 1863, to propose invasion. Lee dissembled as to his real purpose, telling the political leaders that he wanted to tap the abundant food supplies of the North. The cabinet agreed that Lee could cross the Potomac and "threaten Washington, Baltimore, and Philadelphia."[3]

To Davis, Lee promoted a political line to convince the people of the

North that the South would return to the Union if acceptable conditions were met. "When peace is proposed to us," Lee wrote, "it will be time enough to discuss its terms."[4] Such a plan might gain some support in the North, especially among a growing number of people seeking peace. But it nevertheless depended upon Lee's winning a victory over the Union army. If the South lost a battle on northern soil and had to retreat, the peace initiative would collapse.

Lee reorganized his army, creating three corps instead of the two previously. For the new corps, he took one division from Longstreet's corps, another from Jackson's, and formed a third division from two brigades on detached duty and two brigades from A. P. Hill's six-brigade division. Longstreet had returned from the Suffolk campaign to resume command of the 1st Corps. Lee promoted Richard S. Ewell to lead Jackson's old 2nd Corps. Ewell had lost his left leg at the battle of Groveton but now could walk on his wooden leg and could mount and ride a horse. Lee named A. P. Hill as commander of the new 3rd Corps. Ewell and Hill were promoted to lieutenant general.[5]

With Jackson gone, Longstreet became Lee's closest military confidant. Longstreet had absorbed some of Jackson's understanding that attacks against strongly defended Federal emplacements would lead to disaster. He attempted to get Lee to accept this concept as well. The proper strategy, he held, was to give battle only when Confederate forces were in a strong position and ready to receive attack. "Our numbers were less than the Federal forces," he later wrote, "and our resources were limited while theirs were not. The time had come when it was imperative that the skill of generals and the strategy and tactics of war should take the place of muscle against muscle."[6]

At Fredericksburg, he reminded Lee, 5,000 Confederates at Marye's Heights had beaten two-thirds of the entire Federal army. Jackson's idea of defending, not attacking, was now widespread in the Confederate army. Porter Alexander, Longstreet's artillery chief, remarked: "When all our corps were together, what could successfully attack us?"[7]

Longstreet thought he got Lee to promise that defense—not offense— was to be the keynote of the campaign into the North.

✦   ✦   ✦

THE UNION COMMANDER, Joseph Hooker, his army still at Falmouth facing Fredericksburg, became suspicious of Lee's intentions. His spies reported that on June 3 Longstreet's corps began moving to Culpeper, northwest of Chancellorsville. The next day Ewell's corps followed. To find out if this indicated a general movement of the whole army, Hooker sent his cavalry across the Rappahannock River toward Culpeper on June 9.

The Union cavalry, under Alfred Pleasonton, collided with Jeb Stuart's horsemen at Brandy Station, on the Orange and Alexandria Railroad, about six miles northeast of Culpeper. For hours, both sides contested Fleetwood Hill, a long ridge just north of the station, each side charging back and forth. It was the greatest cavalry action of the war. Its only accomplishment, however, was to prove to Hooker that Lee indeed was moving northward. Union General Abner Doubleday explained why pure cavalry fights were indecisive: "Every cavalry charge, unless supported by artillery or infantry, is necessarily repulsed by a countercharge; for when the force of the attack is spent, the men who make it are always more or less scattered, and therefore unable to contend against the impetus of a fresh line of troops, who come against them at full speed and strike in mass."[8]

With Longstreet's and Ewell's corps at Culpeper, Hooker saw that Hill's corps was all alone at Fredericksburg. He proposed on June 10 that he cross the Rappahannock, overwhelm Hill, then strike directly for Richmond, fifty-five miles south. This was a feasible strategy if carried out swiftly. A force of more than 15,000 Union troops at Suffolk, on the Peninsula around Williamsburg, and at White House on the Pamunkey River northeast of Richmond could have cooperated with Hooker's thrust on Richmond. But Hooker's instructions were always to keep the safety of Washington foremost in his planning, and Lincoln rejected the Richmond plan out of hand. "If you had Richmond invested today," he wrote, "you would not be able to take it in twenty days. . . . Lee's army, and not Richmond, is your sure objective point."[9] Soon afterward, Union forces at White House made gestures to attack Richmond, but President

Davis called out militia, and the Union effort disintegrated from lack of resolve and leadership.

On June 10, 1863, Lee set in motion his second invasion of the North. Leading the advance was Ewell's corps, which crossed the Blue Ridge and marched toward Winchester, Virginia. Longstreet's corps marched parallel with Ewell but east of the Blue Ridge. This was in part to shield A. P. Hill, still at Fredericksburg, but also in the hope of goading Hooker into attacking the Confederate army.

On June 13 and 14, Ewell surrounded Union General Robert H. Milroy's garrison at Winchester. Milroy attempted to get away, but a Confederate force blocked his path, and only a small part of the command slipped through. The remainder, about 4,000 men, surrendered. Now there was no force to challenge the road to the North.

Following orders from Lincoln, Hooker turned his army toward Washington on June 13. By June 15 most of his force was around Centreville, Manassas, and Fairfax Court House. A. P. Hill by this point had broken camp at Fredericksburg and marched toward the Shenandoah Valley, and Longstreet had moved from Culpeper north to Ashby's and Snicker's Gaps in the Blue Ridge.[10] Lee hoped to entice Hooker into driving against these gaps, giving the Confederates a chance to crush the Union army in open country or get between Hooker and Washington.

But Hooker would not take the bait. He remained guarding Washington, even though Pleasonton's cavalry had pushed Stuart's horsemen back to the gaps in a series of running fights around Middleburg, Upperville, Aldie, and Philomont. Lee directed Stuart to hold the gaps until the Rebel advance across the Potomac drew Hooker's forces after it. Lee told Stuart that his primary job was to keep the Federals as far east as possible, protect the lines of communications, and scout out enemy movements.[11]

Unable to induce Hooker to pursue him, Lee ordered Hill's corps to follow Ewell and drew Longstreet's corps into the Valley to guard the Confederate rear. Two cavalry brigades went with Ewell's advanced force: John D. Imboden's 2,100-man outfit was to guard the left or west, and Albert G. Jenkins's 1,600-man unit was to stay in front of the Confederate

The Gettysburg Campaign
June 10–July 14, 1863

0 Miles 10 20
0 Kilometers 20

BLUE MOUNTAIN
Conedogwinet Creek
Kingston
Newville
Carlisle
Yellow Breechs
Oakville
Petersburg
CUMBERLAND VALLEY R.R.
KITTATINNY MOUNTAIN
Shippensburg
CUMBERLAND
Bendersville
Heidlersburg
Conococheague Creek
VALLEY
Chambersburg
Mummasburg
Favetteville
Cashtown
New Oxford
New Guilford
SOUTH MOUNTAINS
Mercersburg
Gettysburg
W. Branch
Greencastle
Fairfield
Littlestown
COVE MOUNTAIN
Waynesboro
Marsh
PENNSYLVANIA
MARYLAND
Emmitsburg
Creek
Taneytown
Leitersburgh
BALTIMORE AND OHIO R.R.
NORTH MOUNTAIN
Hagerstown
Cavetown
Pipe Creek
Funkstown
Middleburg
Uniontown
Williamsport
Monocacy
Creek
Catoctin
SOUTH MOUNTAIN
Lewistown
Woodsboro
WEST
Antietam
Boonsboro
VIRGINIA
Falling Waters
Myersville
(Admitted to the Union
June 20, 1863)
Sharpsburg
Middletown
Martinsburg
New Market
Shepherdstown
Burkittsville
Frederick
AND OHIO R.R.
Opequon Creek
Harpers Ferry
CATOCTIN MOUNTAINS
Damascus
Bolivar Heights
Charles Town
Potomac
Point of Rocks
Laytonsville
WINCHESTER AND
River
POTOMAC R.R.
Barnesville
Brookeville
Winchester
Shenandoah River
BLUE RIDGE MOUNTAINS
White's Ferry
Poolesville
Gaithersburg
Berryville
Purcellville
Leesburg
VIRGINIA
Edwards Ferry
Rockville
Snicker's Gap
Dranesville
ALEXANDRIA, LOUDOUN AND HAMPSHIRE R.R.
Upperville
Washington
Ashby's Gap
Middleburg

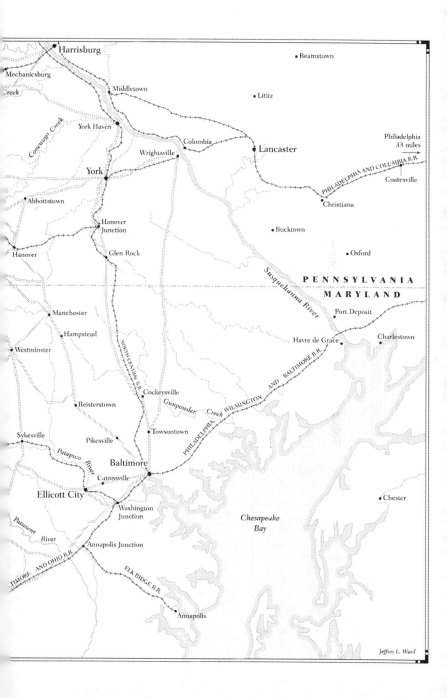

Jeffrey L. Ward

army. Imboden's horsemen broke up parts of the Baltimore and Ohio Railroad, while Jenkins's brigade moved well ahead of Ewell's infantry, arrived at Chambersburg, Pennsylvania, twenty-five miles north of the Potomac, on June 15, and kept going.[12]

Union General Darius N. Couch had been detailed to set up defenses at Harrisburg, Pennsylvania, and along the Susquehanna River. Before he arrived, however, Jenkins's cavalry had already reached Carlisle, eighteen miles west of Harrisburg. Terrified farmers fled across the river, driving cattle and horses before them.

Ewell's infantry arrived at Williamsport, Maryland, on June 15 and began to cross the Potomac River. Edward Johnson's and Robert Rodes's divisions of Ewell's corps reached Chambersburg on June 23. Ewell's other division, under Jubal Early, took the road eastward and halted on June 23 at Waynesboro, Pennsylvania, fifteen miles south of Chambersburg and twenty miles west of Gettysburg.

Governors and civilian leaders in northern states demanded immediate action. Lincoln called on nearby states to raise 120,000 men for temporary service, but few came forward. Even if they had, there was no time to prepare them for war and few additional arms.

Lee worked out his immediate plans. Ewell was to lead the advance northward from Chambersburg to Harrisburg and the Susquehanna and destroy rail communications with the west. Hill was to stay east of South Mountain, the extension into Maryland and Pennsylvania of the Blue Ridge, to keep the enemy at a distance.

Stuart now proposed another grandiose sweep around the Federal army. He asked Lee for permission to leave Imboden's and Jenkins's cavalry brigades to cover Ewell's advance into the North, and Beverly H. Robertson's two brigades to guard Ashby's and Snicker's Gaps through the Blue Ridge. Stuart would take the remainder of his cavalry, 6,000 men, to close up on the rear of Hooker's army. Lee agreed, but enjoined Stuart to rejoin the army at once if Hooker crossed into Maryland. Stuart would need to serve on the Confederate army's right flank as it moved north in order to scout out all Federal movements and report them swiftly to Lee himself.

Lee's decision to allow Stuart's to undertake another of his "rides" was strange and inconsistent. After all, Stuart's scouting out of Hooker's forces had been decisive in making the victory at Chancellorsville possible. It was also illogical to believe that Hooker would be slow to follow Lee into the North and thus give Stuart time to disrupt his rear until the Confederates could capture Harrisburg. Lincoln was certain to require Hooker to press after the Rebel army as soon as possible. He could not allow it to rampage through the North. Lee should have known the Federals would come after him, and soon. Consequently, it was extremely dangerous to allow the Confederate army's entire right flank—this the flank on which the Union army was bound to approach—to be devoid of cavalry reconnaissance and protection for even one day.

Stuart understood his duty, but he also wanted to achieve another spectacular feat of riding entirely around Hooker's army. He started out on June 25, advancing to Haymarket, Virginia, just east of Thoroughfare Gap in the Bull Run Mountains (and also just west of the First and Second Manassas battlefields). There Stuart came upon the Union corps of Winfield S. Hancock. It was evident that the entire Union army was moving north toward the Potomac on all roads. Stuart's clear duty at this moment was to return at once to shield the Confederate army and report the news of the Federal movement to Lee.

Instead, Stuart went on eastward, placing Hooker's army between himself and the Confederate army, and eliminating any chance of scouting for the army. This was a complete failure of responsibility on Stuart's part. Now he had to make a wide detour around the Federal army to get north. He finally found an uncontested crossing of the Potomac near Dranesville, Virginia, fifteen miles upstream from Washington. But by then he was two days behind the Union army, and completely isolated from the Confederate army. Stuart did seize a large wagon train in Maryland, burn a bridge of the Baltimore and Ohio Railroad, and chase a small Union cavalry force out of Westminster, Maryland. But these were small pickings, and he always remained behind the Union army and served no function as Lee's eyes and ears.

There has been much justifiable criticism of Stuart's dereliction of duty. But Lee bears equal blame, not only for allowing Stuart to embark on his ride in the first place but also for his failure—once he realized that

Stuart was not guarding his right flank—to call up Robertson's two cavalry brigades, still idly standing on the Blue Ridge passes long after the Federal army had moved northward. Lee did not summon Robertson until he had learned from one of Longstreet's spies that the Union army had crossed the Potomac and was following him. Robertson arrived so late that he served no function whatsoever. Lee also had at his disposal Imboden's large cavalry brigade, which was useless on the left or western flank. Imboden, as well, could have been called over to the right flank the moment Lee discovered it was bare.

❈   ❈   ❈

FEELING WASHINGTON was safe, Hooker commenced a cautious advance toward Lee. He selected a course on the east side of South Mountain, but parallel to Lee's movements. If Lee turned toward Baltimore or Washington, Hooker believed he could occupy the gaps in South Mountain before Lee could seize them.

The Union army began to cross the Potomac at Edwards Ferry, Virginia, six miles east of Leesburg, on June 25, and marched toward Frederick, Maryland. (Thus the Federals were already in Maryland the day Jeb Stuart began his sweep east.) But Hooker also sent Henry W. Slocum's 12th Corps toward Harpers Ferry to act against Lee's supply line back into the Shenandoah Valley. Lee, realizing that he could not maintain this line, severed it at once. The Army of Northern Virginia now had to live off the country. The army had only enough artillery ammunition to fight one heavy battle. This was enough, however, since a victory would transform the military situation. And Lee believed that, if need be, he could bring up more ammunition by cavalry escort.[13]

Longstreet forded the Potomac at Williamsport on June 24, while Hill crossed at Shepherdstown, passed over the battlefield of Antietam, and united with Longstreet at Hagerstown the next day. On June 27, the two corps arrived at Chambersburg and Fayetteville, eight miles east.

Now supported, Ewell sent Edward Johnson's and Robert Rodes's divisions to Carlisle, which they occupied on June 27, while Jenkins's cavalry struck Kingston, skirmishing only four miles from Harrisburg. Panic gripped Harrisburg, and business halted in Philadelphia.

On June 26 Jubal Early marched through Gettysburg. Passing through this town of three thousand people, Early discovered a shoe factory and sent word back to Hill's corps that a desperately needed supply of footwear was there. Early pressed on to York, then sent a brigade to secure a bridge over the Susquehanna at Wrightsville, twenty miles southeast of Harrisburg, hoping to operate against Harrisburg from the rear. But militia from Columbia, Pennsylvania—just across the river—had set the bridge ablaze.

Hooker decided it was useless to keep a large force at Harpers Ferry and asked General Halleck to add the 10,000-man garrison there to Slocum's corps and to use the combined force to follow on Lee's rear. But Lincoln said the garrison had to remain at the Ferry. Hooker, believing the army should be concentrated, offered his resignation, assuming it would be rejected on the eve of battle.

But Lincoln had been hunting for a way to fire Hooker ever since Chancellorsville. He had been deterred only because Hooker was supported by Salmon P. Chase, secretary of the treasury. Now Lincoln accepted Hooker's resignation on June 28 and replaced him with George Gordon Meade, then commanding the 5th Corps. Meade was a West Pointer, class of 1835, with much experience, but he had attained little success in the war. Meade took command and moved north.

At Chambersburg on June 28, Robert E. Lee ordered Ewell to march from Carlisle to Harrisburg, and Longstreet and Hill to follow, probably the next day. That evening, however, Lee received word from one of Longstreet's spies that three Union corps were around Frederick, with two more near South Mountain, a few miles west, and that Meade was now in command.

Lee reacted in a most odd and destructive way. He decided to concentrate the Confederate army east of South Mountain, in order to keep the Union army out of the Cumberland Valley to the west, thus unable to cut his supply line.[14] But Lee had already abandoned his line of supply back to Virginia and had no rear to protect. At the same time, it was nonsensical to think that Meade might move into the Cumberland Valley. Experience had shown that Lincoln would require the Union army to remain between Lee and Washington, D.C. Meade was certain *not* to get on

Lee's tail, meaning that Lee had no reason to concentrate east of South Mountain.

Lee decided to bring the Confederate army together at Gettysburg. On the night of June 28 couriers rushed off to instruct Ewell to march directly from Carlisle, York, and Wrightsville toward Gettysburg, Hill to move through Cashtown to Gettysburg eight miles east, and Robertson's horsemen, still guarding the Blue Ridge passes, to come forward. Longstreet was to follow the next day, leaving George Pickett's division to guard the army's rear until Imboden's cavalry, two days' journey to the west, could arrive.

Lee's decision to concentrate the army at Gettysburg was senseless. Even without the scouting of Stuart's horsemen, he had attained a superb strategic position by his march into Pennsylvania. His army was well north of the Federal army. Long before Meade could have reacted, Lee could have consolidated his army, crossed the Susquehanna, seized Harrisburg, broken the bridges, turned the river into a moat, and had a long head start down the uncontested road to Philadelphia. This was the Union's second city, with 600,000 people, directly on the main north-south railroad corridor, and absolutely vital to the North. If Lee moved on Philadelphia, Meade would be obliged to stop him if he could. Anywhere along the way where there was a good defensive position with open flanks, Lee could have halted, drawn up his army, and waited. Meade would have been forced to attack. He almost certainly would have been defeated, and the entire war would have been transformed.

Instead, Lee ordered the forced march on Gettysburg. It was like stepping off a cliff in the dark. He had no idea what lay ahead. Lee admitted in his official report that he knew nothing about the size of enemy forces advancing on Gettysburg or what their intentions were.[15] In fact, Union forces were closer to Gettysburg than were the Confederates. On this day three Union corps and John Buford's cavalry division were fifteen miles from Gettysburg. The Rebel forces under A. P. Hill remained twenty miles away, with the exception of Harry Heth's division, which got to Cashtown on June 29. Longstreet and Ewell were twenty-five miles distant, Early's advance elements forty. The Confederates would arrive exhausted at a place that had not been reconnoitered—and, as it turned out, was already occupied by Buford's cavalry when the first Rebels arrived. Furthermore, Lee knew nothing of the town's suitability for defense, nothing of the terrain.[16]

If Lee could have not brought himself to march on Philadelphia—by far the best strategic move—Lee could have ordered the Confederate army to concentrate at Carlisle. The Confederates already knew the terrain around Carlisle, and Meade would take days to reach it. Lee would have had plenty of time to search out a suitable defensive position, build entrenchments, and await the Union army. Although a battlefield around Carlisle would not have forced Meade into an immediate, frantic attack on unexplored ground, as would have been the case if Lee were threatening Philadelphia, Meade nevertheless would have been obligated to attack, and he would have been defeated there as well. Second Manassas and Fredericksburg had proved that frontal attacks against prepared positions were almost bound to fail. The truth would be demonstrated anew at Gettysburg only a few days later. Defense, not attack, was the plan that Lee and Longstreet had agreed upon before the invasion of Pennsylvania had begun.

The showdown with Meade should *never* have taken place at Gettysburg. If Stonewall Jackson had been alive, he might have talked Lee out of his wrong decision, just as he had finally persuaded Lee to embrace his flank attack at Chancellorsville. But Longstreet, who did not have the influence of Jackson, could not.

❋   ❋   ❋

BY NIGHTFALL ON June 30, Robert Rodes's and Jubal Early's divisions of Ewell's corps arrived at Heidlersburg, ten miles north of Gettysburg, while Dorsey Pender's division of Hill's corps was behind Heth's division at Cashtown.

With no cavalry to inform him, Lee didn't know that Federal horsemen were fanning out northward, trying to shield a possible movement of the Federal army toward Harrisburg. On June 29, Buford's 3,000 cavalry troopers passed near Fairfield, six miles south of Cashtown, and saw the bivouac fires of a small Rebel detachment that Johnston Pettigrew, leading Heth's advance brigade, had set out as a flank guard. Buford withdrew south to Emmitsburg, Maryland, and told Union cavalry chief Alfred Pleasonton that the Rebels were advancing from Fairfield. Pleasonton told Buford to get to Gettysburg the next day and hold it until the Federal infantry arrived.

On June 30, Harry Heth at Cashtown sent Pettigrew's brigade on to Gettysburg, hoping to find the shoes that Early had reported were there. Atop a ridge a mile west of the town, Pettigrew saw through his field glasses the long dark column of Buford's cavalry approaching on the road from Emmitsburg. He had orders not to get into a lone fight, so he withdrew about four miles, rode back to Cashtown, and informed Heth and Hill.

After the Confederates had departed, General Buford rode to the top of the hill where Pettigrew had stood and watched the Rebel infantry retire. He informed Meade and General John F. Reynolds, commanding the left wing of the Union army (1st, 3rd, and 11th Corps), that large bodies of Confederates were emerging from the South Mountain pass on the Cashtown Road.[17] This confirmed reports Meade had already received that the Rebels had abandoned the threat against Harrisburg and Philadelphia and were concentrating toward Washington and Baltimore.

When Meade learned that the Confederate flank guard was at Fairfield and Rebels were streaming down the Cashtown Road, he concluded that Lee was coming after him. Any idea of a bold advance vanished. On June 30, Meade ordered the entire Union army to retreat back to Pipe Creek, eighteen to twenty miles southeast of Gettysburg, to take up a defensive position twenty miles long from Middleburg on the west to Manchester on the east. Along Pipe Creek, Meade could protect both Washington and Baltimore—but not Harrisburg or Philadelphia.

Buford, knowing nothing of Meade's decision to retreat, was sure Reynolds would bring up infantry to block the Confederates at Gettysburg the next day, July 1, and determined to hold his ground on the mostly cleared western elevation, known as McPherson Ridge, until Union infantry could arrive. For his part, Reynolds probably never received the order to withdraw to Pipe Creek, and followed Meade's original orders to advance on Gettysburg on July 1.

On the Confederate side, A. P. Hill reasoned from Pettigrew's report that the Federal cavalry were only an observation detachment, not an occupying force. Harry Heth, seeing an opening, asked Hill whether he would have an objection to taking his division to Gettysburg the next day to get those shoes.

"None in the world," Hill replied.[18]

On the night of June 30, Hill informed Lee of Pettigrew's encounter with Federal cavalry at Gettysburg and sent a courier to Ewell advising that he was going to advance on the town the next morning. Harry Heth's division started out early Wednesday, July 1, followed by Dorsey Pender's division. Hill had no positive information as to what lay ahead, and no other purpose than to secure some badly needed shoes. In this purely accidental fashion Hill set off the greatest battle ever to be fought in the Western Hemisphere.[19]

※   ※   ※

ON THE MORNING of July 1, the Federal army was spread out along the Maryland-Pennsylvania border. Closest was the 1st Corps, under Abner Doubleday, on the extreme left, six miles southwest of Gettysburg at Marsh Creek, and the 11th Corps under Oliver O. Howard, a couple of miles south of Emmitsburg. Meade's other five corps were scattered south and southeast for twenty miles. The Army of the Potomac counted 80,000 men in its seven corps, plus 13,000 cavalry. The Army of Northern Virginia had about 60,000 men in its three corps, and 10,000 cavalry, though Stuart was off with most of the horsemen.

When Heth's division approached Gettysburg on the Cashtown Road, otherwise known as the Chambersburg Pike, the infantry pushed Buford's vedettes across Willoughby Run, a mile west of town and at the foot of McPherson Ridge. About 6:00 A.M. Buford sent word of the advance to General Reynolds, who was at Marsh Creek. Reynolds ordered both the 1st Corps and 11th Corps to march at once for Gettysburg. Reynolds himself went with the 1st Division, under James S. Wadsworth, while Abner Doubleday brought up the other two divisions.

Heth attacked straight ahead, not waiting for Pender to come up and

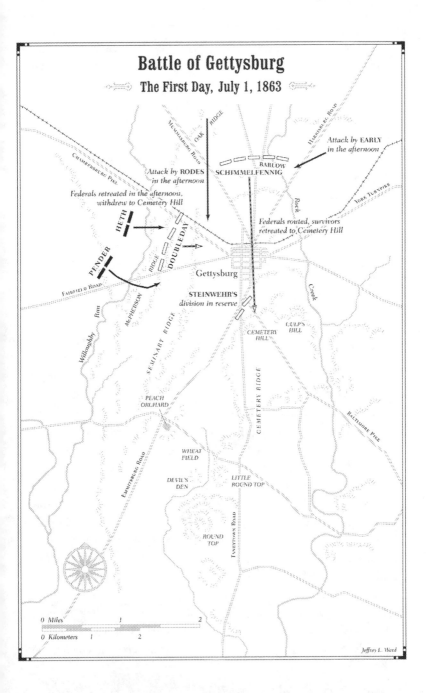

Battle of Gettysburg
The First Day, July 1, 1863

MUMMASBURG ROAD

OAK RIDGE

CHAMBERSBURG PIKE

HARRISBURG ROAD

Attack by EARLY
in the afternoon

BARLOW

SCHIMMELFENNIG

Attack by RODES
in the afternoon

Rock

YORK TURNPIKE

Federals retreated in the afternoon,
withdrew to Cemetery Hill

HETH

DOUBLEDAY

Federals routed, survivors
retreated to Cemetery Hill

PENDER

RIDGE

FAIRFIELD ROAD

Gettysburg

Run

Creek

McPHERSON

STEINWEHR'S
division in reserve

SEMINARY RIDGE

Willoughby

CEMETERY
HILL

CULP'S
HILL

CEMETERY RIDGE

BALTIMORE PIKE

PEACH
ORCHARD

EMMITSBURG ROAD

WHEAT
FIELD

DEVIL'S
DEN

LITTLE
ROUND TOP

TANEYTOWN ROAD

ROUND
TOP

0 Miles   1   2

0 Kilometers   1   2

Jeffrey L. Ward

not sending back for instructions. But he moved with extreme slowness, giving Buford time to organize a defense with his dismounted cavalry until Wadsworth's infantry arrived and occupied McPherson Ridge.

Heth now sent in two unsupported attacks—Joseph R. Davis's brigade on both sides of the Cashtown Road and James J. Archer's brigade about a quarter of a mile south. Davis drove three Union regiments back into Gettysburg but allowed two of his regiments to get cornered in a deep railroad cut just north of the road, where hundreds surrendered. Archer was careless and permitted the Federals' Iron Brigade to sweep around his flank and capture nearly a hundred Rebels, including Archer himself.

Harry Heth had been told explicitly not to bring on a general engagement, but this was precisely what he was doing. Moreover, he was challenging an enemy of unknown strength without informing his superiors or calling for help.

Accordingly, the billowing battle came as a complete surprise both to Robert E. Lee and to A. P. Hill. Lee rode to the crest of South Mountain, heard the distant rumble of artillery, and hurried down to Cashtown at the foot of the mountain. There he encountered Hill, who knew only that Heth and Pender had gone ahead with orders not to force an action. Hill rode off to find out what was happening.

Lee was waiting impatiently for word from Hill when General Richard Anderson arrived. If the force ahead was the whole Federal army, Lee told Anderson, "we must fight a battle here."[20] Tired of waiting for Hill to report, Lee rode on toward Gettysburg, arriving on the field about 2:00 P.M. He found that the Confederates were responding to the battle as it developed, had no idea what Federal force lay ahead, and had no plan except to drive straight ahead.

Shortly after Union General Reynolds arrived on McPherson Ridge, he was struck dead by a Rebel sharpshooter. Oliver O. Howard, chief of 11th Corps, who had arrived on the field, took command. Meade, back at Taneytown fifteen miles south, had little confidence in Howard after his poor performance at Chancellorsville, so he called on Winfield S. Hancock, chief of the 2nd Corps at Taneytown, to take command of the left wing and to go at once to Gettysburg. Hancock, though junior to a number of other Union generals, was highly regarded in the army. Meade told Hancock he wanted to move the forces at Gettysburg back to Pipe Creek. But if the ground at Gettysburg was suitable and circumstances made it wise, he said, Hancock had the authority to establish a line of battle at Gettysburg.

Before Hancock could get to Gettysburg, Ewell's corps, coming from the north, occupied Oak Ridge, an elevation north of McPherson Ridge, and was threatening to advance directly onto the flank of the Federals, who now were being directed by General Doubleday. Doubleday told Howard that he could hold off Heth to the west if Howard could stop Ewell from attacking his right from Oak Ridge.

Howard moved two of his divisions—Francis C. Barlow's and Alexander Schimmelpfennig's—into the open, cultivated country just north of town. Barlow's right rested where the road from Harrisburg (now U.S. Route 15 Business) crossed Rock Creek. Schimmelpfennig lined up to the west of Barlow but did not reach all the way to Oak Ridge, leaving half a mile of undefended space between Doubleday's and his positions. Both ends of Howard's north-facing line thus were open, leaving them vulnerable to a flank attack by Ewell.[21] As a reserve, Howard moved his third division, under Adolph von Steinwehr, to Cemetery Hill, which rose a hundred feet above the countryside, just south of Gettysburg.

Confederate General Robert Rodes, seeing that the Federals had left a huge gap between their two corps, sent three North Carolina brigades south from Oak Ridge. They evicted Doubleday's reserves but suffered heavy casualties; a desperate Federal stand at the Cashtown Road by Colonel Roy Stone's Pennsylvania brigade finally stopped Rodes's advance.

From west of McPherson Ridge, Heth watched Rodes's attack as it developed. Riding back to Lee, he said: "Rodes is heavily engaged. Had I not better attack?"

"No," Lee answered, "I am not prepared to bring on a general engagement today—Longstreet is not up."

Heth returned to the line and watched Rodes's difficulties on the Cashtown Road. He went back to Lee and asked permission again to attack. This time Lee agreed.[22] With no purpose other than to help Rodes in his fight on the Cashtown Road, Lee abandoned the defensive policy he had agreed to with Longstreet. The sound of the guns once more had inflamed Lee's inherent combativeness. Here at Gettysburg on July 1, 1863, he did as he had done at Beaver Dam Creek, Frayser's Farm, and Malvern Hill in the Seven Days—he abandoned his larger strategic aims and turned directly on the enemy the moment he was confronted with a surprise or a challenge.

By allowing himself to be drawn into an offensive battle, Lee passed up the opportunity to pull back his entire army to find out Meade's intentions and locate a suitable battlefield in case the Federals attacked. Such a position was readily available—on the ridgeline just above Cashtown, only eight miles to the west. If Meade had attacked there, he would have been defeated. In reality, Meade most likely would not have attacked but would have retreated to Pipe Creek, placing himself on the defensive and opening up immense opportunities for the Confederate army, which could have struck out for Philadelphia facing no opposition.

Instead, Lee ordered Heth to make a frontal assault directly into Doubleday's position, while Dorsey Pender's division, which had arrived, advanced on the south, overlapping Doubleday's line for a quarter of a mile. A bullet struck Heth in the head almost at once. Though his hat absorbed much of the blow, Heth remained insensible for the next day and a half. Pettigrew took command.

General A. P. Hill took little part in the movements of his corps. British Lieutenant Colonel Arthur J. L. Fremantle, accompanying the Confederate army as an observer, wrote that Hill reported he had been unwell all day.[23]

The principal fight took place between Pettigrew's North Carolina brigade and the Iron Brigade on McPherson Ridge. The Tar Heels marched straight into a hail of canister and rifle fire but forced the Iron Brigade back to a reserve line. There the North Carolinians' final, desperate charge broke the Federal resistance and sent the survivors fleeing to the rear. Doubleday ordered the 1st Corps back to a last stand around the Lutheran Theological Seminary, on the mostly wooded Seminary Ridge a half mile east of McPherson Ridge.

The losses on both sides were appalling. Doubleday lost 5,000 men, half the corps' strength. Heth's force, in less than half an hour, lost 2,700 of 7,000 men. The 26th North Carolina suffered 80 percent casualties. The 24th Michigan, which it challenged, lost 399 out of 496 men.

As soon as Ewell saw that the Federals were closely engaged on McPherson Ridge, he launched Jubal Early's division east of the Harrisburg Road onto the right flank of the 11th Corps, enveloping Francis Barlow's position. Both Barlow's and Schimmelpfennig's divisions collapsed; many men surrendered, and others ran back into town, only to be captured by Ewell's troops.

Ewell's seizure of the town threatened to cut off 1st Corps at the seminary, and around 4:00 P.M. Howard ordered the survivors of both the 1st Corps and 11th Corps to rally on Cemetery Hill, where Steinwehr's division was still emplaced.

Cemetery Hill was at the northern end of Cemetery Ridge, an elevation that stretched southward in an arc from the burial ground for two miles to Little Round Top, 480 feet high, with considerable open ground and much exposed rock, and heavily wooded Round Top, 600 feet high, just to the south. Half a mile east of the cemetery, forming sort of a fishhook, was Culp's Hill, about a hundred feet higher than Cemetery Hill. Paralleling the ridgeline about a mile to the west across open, mostly cultivated fields was Seminary Ridge, now occupied by the Confederates. Between the two ridges ran the Emmitsburg Road (now U.S. Route 15 Business).

Confederate Porter Alexander, who knew it well, called Cemetery Ridge "the most beautiful position for an army which I have ever seen occupied. Good positions were abundant in this section, it being remarkably well cultivated and having numerous extensive ridges, with open rolling lands between."[24]

Union General Winfield S. Hancock arrived on Cemetery Hill about the time the Federal soldiers were fleeing to it. Looking down Cemetery Ridge, Hancock told General Howard: "I think this is the strongest position by nature upon which to fight a battle that I ever saw." The issue, he felt, should be decided here, not back at Pipe Creek, and at once he sent a courier to Meade proposing that the Union army concentrate at Gettysburg. Meade accepted Hancock's recommendation, sent orders for all corps to march at once, and got to Gettysburg himself around 11:30 P.M. on July 1. Of the seven Union corps, only John Sedgwick's 6th was not nearby shortly after dark. It was still at Manchester, Maryland, nearly thirty miles away, but began marching in the early hours of July 2.

Meade had no idea of attacking, only of defending Cemetery Hill and Ridge. Hancock set to work to secure the site, ordering up cannons, placing soldiers behind stone fences, and directing James Wadsworth's division to occupy Culp's Hill.

*   *   *

GENERAL JAMES LONGSTREET arrived on Seminary Ridge after the fighting had ended on July 1. He found Lee on the summit watching the Federals concentrate on Cemetery Hill.

Longstreet studied the situation with his glasses for five or ten minutes. Then he turned to Lee. "All we have to do," he said, "is to throw our army around by their left [that is, to the south], and we shall interpose between the Federal army and Washington." There the Confederates could get into a strong position and wait. If the enemy failed to attack, the Confederates could move back toward Washington and select a good position to receive battle. "The Federals will be sure to attack us," Longstreet said. "When they attack, we shall beat them."

"No," General Lee said, "the enemy is there [pointing to Cemetery Hill], and I am going to attack him there."

Longstreet argued hard against Lee's resolve. Moving to the south would give the Confederates control of the roads leading to Washington and Baltimore, he pointed out. If the Confederates got between the Federal army and Washington, Meade would be compelled to attack and would be badly beaten.

Again Lee refused. "No," he said, "they are there in position, and I am going to whip them or they are going to whip me."[25]

Lee's mind was made up. He was going to assault the Union army head-on. Forgotten was all the resolve he had made to stand on the defensive. Ignored was Napoléon Bonaparte's injunction never to make a frontal attack if one could do otherwise, and always to move on the enemy's rear, even if it failed, because a rear maneuver would shake enemy morale and force him into a mistake.[26] And disregarded was Longstreet's entirely correct assessment of the situation—that Meade's army could be turned on its southern flank with little difficulty.[27] (Union General Abner Doubleday confirmed this when he later wrote, "Lee could easily have maneuvered Meade out of his strong position on the heights, and should have done so.")[28] Instead, Lee fell back into the pattern he had demonstrated earlier this same day, and had exhibited since he first took command of the army in June 1862.[29]

For the South, the saddest part of Lee's decision was that even if he could not have brought himself to move around Meade's southern flank, he *still* could have won simply by remaining stationary on Seminary Ridge. Though Meade's initial thought was only to defend Cemetery Ridge, in fact he would have been obliged to go on the attack if Lee did not launch an assault. By remaining idle, Meade would have not only angered Abraham Lincoln and the Northern populace but also left Lee free to move off north, south, or west to extract more goods from Northern farmers and factories and destroy more Northern railroads. No, if Lee had not attacked, Meade would have been forced to do so.

Porter Alexander described the situation beautifully. "It does not seem improbable that we could have faced Meade safely on the 2nd [of July] at Gettysburg without assaulting him in his wonderfully strong position. We had the prestige of victory with us, having chased him off the field and through the town. We had a fine defensive position on Seminary Ridge ready at our hand to occupy. It was not such a really *wonderful* position as the enemy happened to fall into, but it was no bad one, and it could never have been successfully assaulted." Alexander added that "the onus of attack was upon Meade anyhow. We could even have fallen back to Cashtown and held the mountain passes with all the prestige of victory, and popular sentiment would have forced Meade to take the aggressive."[30]

On the late afternoon of July 1, 1863, the South came to a crossroads. In one direction lay victory; in the other, defeat. Lee rejected the path to victory that James Longstreet pointed out clearly, and defiantly insisted on the path of direct confrontation, which could only spell the end of Southern independence.

* * *

LEE HAD DECIDED to attack on the second day, July 2, but he had not decided where. During the evening he met with Ewell, Early, and Rodes. He wanted Ewell to seize Cemetery Hill and Culp's Hill early the next morning. Early, speaking for Ewell, argued that Federal forces were concentrating on these heights and that the best chance for success lay in seizing the Round Tops at the south end of Cemetery Ridge. These elevations dominated the field. If they were seized, the Federals would be compelled to evacuate all of Cemetery Ridge, Cemetery Hill, and Culp's Hill.

Lee agreed, and proposed to bring Ewell's entire corps over to Seminary Ridge. This would greatly reduce the length of the Confederate line, which was concave in shape and extended five miles from Culp's Hill around to a point facing Round Top. The Union line, in contrast, was convex and stretched a little more than three miles on high ground, thereby permitting rapid transfer of troops to any point of danger. In addition, the Union army was 25 percent larger than the Confederate army.

Early and Ewell urged keeping the corps in place, assuring Lee that the enemy would not break through. This was not the point. Rather, the key benefit Lee would gain by moving Ewell's corps would be to reduce the tremendous advantage the Federals had with their shorter exterior lines: the Union army could place 25,000 infantry and a hundred guns per mile, whereas the Confederates could deploy only 13,000 infantry and fifty guns per mile.

But when Ewell said he thought he could seize Culp's Hill early the next morning, Lee made the mistake of granting him permission to stay in place and try to take the hill. It turned out that Ewell would not be able to take the position the next morning. In fact, he would contribute little to the battle. Worse, by leaving Ewell's corps on the north, Lee in effect neutralized one-third of the Rebel army and kept his exterior lines dangerously long.

When Lee got back to Seminary Ridge, he learned that Stuart, at last, had materialized. He was at Carlisle, was hurrying to the battlefield, but couldn't arrive until late on July 2. Stuart had ridden over a large part of southern Pennsylvania searching for Confederate forces.

❋   ❋   ❋

THE SECOND DAY of Gettysburg, July 2, 1863, involved three major actions. The largest by far was Longstreet's attack in the south. He drove back a salient that Daniel Sickles's 3rd Corps had formed at the Peach Orchard on the Emmitsburg Road, but he failed to capture Little Round Top. The effect of the incredibly costly engagement was merely to drive Sickles back to Cemetery Ridge, the *real* Union defensive line. Longstreet thus rectified a line that Sickles had overextended in the first place by advancing out to the Peach Orchard. The second fight on July 2 was by Ewell on Culp's Hill. But Meade attacked Ewell at daylight and wrenched from him the ground commanding Culp's. Ewell attacked again at dusk but accomplished nothing. The third event of the day occurred when Ambrose R. Wright's Georgia brigade seized the center of Cemetery Ridge. The breakthrough could have ended the war right there with a great Confederate victory, but Lee had made no preparations for a success in the middle of the line, and A. P. Hill exerted no effort to reinforce Wright. Thus the Georgians were forced to retreat.

Around 5:00 A.M. on July 2 Longstreet met Lee near the seminary building and again proposed to move on Meade's left flank. Again Lee rejected the plan, and ordered the attack on the south end of the Union line. But, ignoring Ewell's and Early's recommendation, he did not select the two Round Tops as his objective. Instead he told Longstreet to seize the lower reaches of Cemetery Ridge and drive the Federals north along it. That is, he was to move up the Emmitsburg Road from the south and strike Cemetery Ridge from the southwest.

Lee believed that only two Federal corps were occupying the heights and that they were facing mainly north on Cemetery and Culp's hills. For this reason, he concluded that he could seize Cemetery Ridge from the south with only two of Longstreet's divisions—John Bell Hood's and Lafayette McLaws's. In fact, Meade sent three times as many Federal troops to the south as were contained in Longstreet's two divisions. The Union general emplaced the 1st, 11th, and 12th Corps on Cemetery and Culp's hills, and Hancock's 2nd Corps and Daniel E. Sickles's 3rd Corps down Cemetery Ridge. George Sykes's 5th Corps went into reserve on

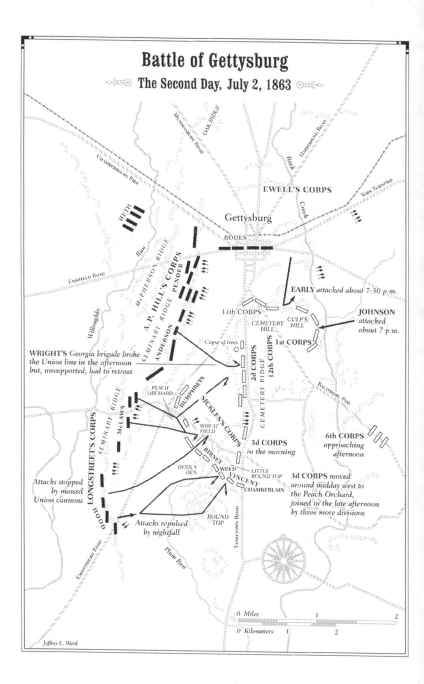

Battle of Gettysburg
The Second Day, July 2, 1863

CHAMBERSBURG PIKE

MUMMASBURG ROAD

OAK RIDGE

HARRISBURG ROAD

Rock Creek

YORK TURNPIKE

EWELL'S CORPS

HETH

Gettysburg

FAIRFIELD ROAD

RODES

McPHERSON RIDGE

A. P. HILL'S CORPS

SEMINARY RIDGE

PENDER

ANDERSON

Willoughby Run

EARLY *attacked about 7:30 p.m.*

11th CORPS

CEMETERY HILL

CULP'S HILL

JOHNSON *attacked about 7 p.m.*

1st CORPS

Copse of trees

2d CORPS

12th CORPS

WRIGHT'S *Georgia brigade broke the Union line in the afternoon but, unsupported, had to retreat*

PEACH ORCHARD

HUMPHREYS

SICKLES'S CORPS

CEMETERY RIDGE

BALTIMORE PIKE

McLAWS

SEMINARY RIDGE

WHEAT FIELD

BIRNEY

3d CORPS *in the morning*

6th CORPS *approaching afternoon*

DEVIL'S DEN

WEED

VINCENT

LITTLE ROUND TOP

CHAMBERLAIN

Attacks stopped by massed Union cannons

LONGSTREET'S CORPS

EMMITSBURG ROAD

3d CORPS *moved around midday west to the Peach Orchard, joined in the late afternoon by three more divisions*

ROUND TOP

TANEYTOWN ROAD

Attacks repulsed by nightfall

Plum Run

0 Miles   1   2

0 Kilometers   1   2

Jeffrey L. Ward

the Baltimore Pike behind. Sedgwick's 6th Corps, when it came up, was to be posted in the rear of the Round Tops as a general reserve.

Longstreet's first objective was the Peach Orchard, on the Emmitsburg Road a mile northwest of Little Round Top. The orchard covered a small rise halfway between Cemetery and Seminary ridges. On it, Confederate artillery could cover an attack on Cemetery Ridge.

Ewell was to strike on the north as soon as he heard Longstreet's cannons, while Hill was to engage the enemy directly ahead on Cemetery Ridge to keep Meade from sending additional troops against Longstreet. Hill was also to attack if he saw the opportunity. In the event, neither Ewell nor Hill would exhibit much initiative.

Longstreet, who greatly opposed this plan, took a long time getting troops into position, waiting until 1:00 P.M. to begin moving his troops. Meanwhile, General Meade ordered Sickles to deploy his two-division 3rd Corps down Cemetery Ridge to Little Round Top, "provided it is practicable to occupy it." Sickles, however, decided it would be better to move west to the high ground around the Peach Orchard, where his artillery could be better deployed. He also saw that the terrain directly in front of the Round Tops would afford excellent cover for enemy troops because it was broken and full of trees and granite outcrops, deserving its name, Devil's Den.

Meade was not especially concerned, because he was convinced the Confederate attack would come on the north. Even so, he sent his artillery chief, General Henry J. Hunt, to scout out the Peach Orchard. Hunt saw that the orchard formed a salient into the Confederate front and could be hit with enfilade artillery fire from the sides, so he refused to authorize it. Nevertheless, Sickles moved his 10,000 infantry forward, making the Peach Orchard the westernmost point of an arc, with David B. Birney's division spreading out thinly to the southeast and Andrew A. Humphrey's division emplacing along Emmitsburg Road to the northeast. Sickles's line was a mile and a half long, twice the distance he had been told to defend on Cemetery Ridge. The southernmost anchor of Birney's line was the Devil's Den. He had no troops to occupy either of the Round Tops.

It was about 3:30 P.M. before Longstreet realized that Sickles had occupied the Peach Orchard, nullifying his plan to locate his cannons on it. Longstreet was forced to form his line of battle with McLaws facing the Peach Orchard and Hood deployed well to the south, opposite Round Top, with E. McIvor Law's Alabama brigade at the southern end, just east of the Emmitsburg Road.

No longer was it possible to seize Cemetery Ridge quickly and roll up the Federal line from south to north. The Confederates would have to break Sickles's advance line first, weakening the blow they could make against the ridge. Longstreet did not inform Lee of the news or ask him to reconsider his plan. He stubbornly held to Lee's original concept, though he knew it no longer could be carried out.[31]

In the midst of this, McIvor Law saw a way out: he could strike straight for Round Top, which his scouts had found unoccupied. Law later recalled, "There now remained not a shadow of doubt that our true *point d'appui* [base] was Round Top, from which the right wing could be extended toward the Taneytown and Baltimore roads [to the east], on the Federal left and rear."[32]

Law told Hood that his brigade could seize Round Top, move around its southern flank to the Taneytown Road, a quarter mile away, and then head to the Baltimore Pike, a couple of miles beyond that. This would unhinge the entire Union position on Cemetery Hill and Ridge. Hood agreed but said his orders were to attack straight ahead. Hood sent word to Longstreet endorsing Law's proposal. Longstreet replied that Lee was already fretting over the delay in attacking and that he was unwilling to offer any further suggestions. Soon afterward, Longstreet rode up and said: "We must obey the orders of General Lee."[33]

Longstreet should have launched both McLaws and Hood simultaneously in a converging attack on the Peach Orchard salient. This would have been the shortest line of attack, could have destroyed a large part of Sickles's corps, and might have allowed the Confederates to reach the southern part of Cemetery Ridge before other Federals could arrive in strength.

But Longstreet launched Hood first and alone—straight for the

Devil's Den and the Round Tops, provoking a wild and chaotic struggle. Hood was wounded quickly, and Law took over as commander. The 47th and 15th Alabama regiments, under Colonel William C. Oates, reached the summit of Round Top with little difficulty, but efforts to seize Little Round Top failed, with heavy casualties.[34] The decision finally came down to a collision between the Alabamians and the 20th Maine, commanded by Colonel Joshua L. Chamberlain, a former professor at Bowdoin College. The Alabama regiments came charging up the rear or south slope of Little Round Top right at the 20th Maine shortly after it took up a defensive position. After a bloody struggle, the Federals drove the Alabamians down the hill and ensured Federal possession of Little Round Top. Chamberlain lost 200 of his 386 soldiers, the Alabamians about the same number.[35] During the night, the Federals, at last realizing Round Top's importance, occupied it in force.

Longstreet waited an hour and a half before sending McLaws into the attack on the Peach Orchard at 6:30 P.M., just as the struggle for Little Round Top was reaching its zenith. After fierce and sustained assaults, the Confederates at last cracked the center of Sickles's line at the orchard, forcing Sickles's entire corps, plus reinforcements, to withdraw, in increasing disorder. The Rebels were stopped by darkness and a gun line established by Federal cannoneers along Plum Run, just west of Cemetery Ridge.

Meade had sent 50,000 soldiers to oppose the Confederates at the southern end of the line. He was left with only a single brigade on Culp's Hill, and few troops elsewhere. Thus a sustained attack by Ewell and Hill might have carried Cemetery Ridge. But Ewell didn't move in the north, and Hill fired his artillery only in support of Longstreet. At the end of the day, though, Hill sent two brigades to assist in driving Andrew Humphreys's Union division back to Cemetery Ridge.

About this time, Ambrose R. "Rans" Wright's 1,800-man Georgia brigade advanced on the center of Cemetery Ridge. Because Union defenders had been thinned to assist in the fight to the south, Wright seized a spot about half a mile south of Cemetery Hill. By the time the Georgians reached the crest, however, they had lost many men, and their impetus. Without reinforcements, the Georgians were forced to withdraw when Meade and Abner Doubleday sent forces streaming from north and south against the Rebel penetration.[36]

Hill failed miserably, but so did Ewell. He did not attack until almost nightfall. One division got up on Culp's Hill but stopped in the darkness and was ripe to be evicted the next morning. The ultimate failure was Lee's. He did not require Hill and Ewell to launch a concerted drive on the Union army. Everything was piecemeal and everything failed.

Although Hood's and McLaws's divisions had lost nearly 4,000 men in this gigantic battle, all the Confederates had done was to drive the Federals back on their main line of defense, Cemetery Ridge. At the end of the day the Union army was in a stronger position than it had been at the outset.

* * *

ROBERT E. LEE's combative personality made it nearly impossible for him to conduct a campaign that was offensive in strategy and defensive in tactics. Seeing war as a direct challenge of one side against the other, he was little interested in maneuvering to place the enemy in an impossible strategic or tactical situation. In this he was the opposite of Stonewall Jackson. Using maneuver alone, Jackson—by descending on Manassas Junction in August 1862—had forced John Pope to abandon his defensive line along the Rappahannock. Again using maneuver alone, Jackson had convinced Pope that the Confederates were cowering in fright at Groveton and thereby induced him to launch one forlorn assault after another against Jackson's defenses.

Lee's proper course was to maneuver so as to threaten at least one objective that General Meade could not permit him to secure. There were three such objectives—Washington, Baltimore, and Philadelphia. Possession of any of them would end the war in a Southern victory.[37] Lee therefore had only to select the objective, take up a defensive position between it and the Federal army, and wait. Meade would be forced to attack. Given the conditions of warfare in 1863, an attack on a previously selected and prepared position was almost bound to fail.

Because of Lee's faulty reasoning, the advantages the South had gained by invading the North had been washed away by the evening of

July 2, 1863. The only reasonable course remaining was to withdraw back into Virginia while the Confederate army still possessed some of its offensive power. But this Lee refused to do. Instead, he insisted on launching a headlong assault against the very center of the Union position—an action that has gone down in history as Pickett's Charge. Of all the counterproductive actions Lee took in the war, this was by far the most damaging.

Lee ordered the assault early on the morning of July 3, when he directed Longstreet to attack into the center of the Union line between Cemetery Hill and Little Round Top. This was the strongest part of the Federal position. Union artillery there had an unobstructed field of fire to the front, and from Cemetery Hill on the north and Little Round Top on the south, cannons could enfilade any line of troops attacking eastward. Moreover, Meade could easily reinforce the center from either side.

Longstreet protested, telling Lee he had reconnoitered the ground out to the south and found that the army could still move in that direction. Longstreet still hoped that he could convince Lee to move around the Federal left.

"No," Lee replied, "I am going to take them where they are on Cemetery Hill. I want you to take Pickett's division and make the attack. I will reinforce you by two divisions of 3rd Corps." These divisions were Heth's, now under Pettigrew, and Pender's, now under Isaac Trimble, Pender having been mortally wounded on July 2. Lee also ordered two brigades of Richard H. Anderson's division—Cadmus M. Wilcox's and David Lang's—to come behind on Pickett's right rear, to guard against an attack on his flank.[38]

Lee was irritated when Longstreet told him that no such body of men "could make that attack successfully." When Longstreet argued that Union guns on Little Round Top could enfilade the attacking columns, Lee called on an artillerist from his staff, Colonel Armistead Long. The colonel said the guns could be suppressed. Porter Alexander, an artillery officer of renown, wrote later: "It seems remarkable that the assumption of Colonel Long so easily passed unchallenged that Confederate guns in open and inferior [that is, lower] positions could 'suppress' Federal artillery fortified upon commanding ridges." Alexander also pointed out

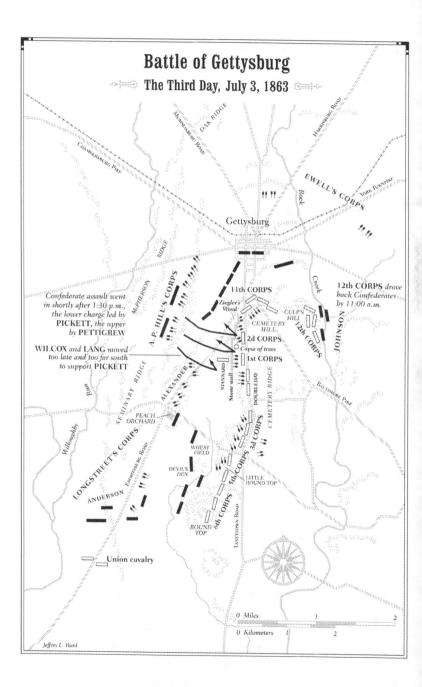

Battle of Gettysburg
⇒⇒⇒ The Third Day, July 3, 1863 ⇐⇐⇐

Confederate assault went in shortly after 1:30 p.m., the lower charge led by **PICKETT,** *the upper by* **PETTIGREW**

WILCOX *and* **LANG** *moved too late and too far south to support* **PICKETT**

12th CORPS *drove back Confederates by 11:00 a.m.*

Gettysburg

EWELL'S CORPS

A. P. HILL'S CORPS

11th CORPS

Ziegler's Wood

CULP'S HILL

CEMETERY HILL

JOHNSON

12th CORPS

2d CORPS

Copse of trees

1st CORPS

STANNARD

Stone wall

DOUBLEDAY

CEMETERY RIDGE

ALEXANDER

LONGSTREET'S CORPS

PEACH ORCHARD

BALTIMORE PIKE

SEMINARY RIDGE

WHEAT FIELD

3d CORPS

DEVIL'S DEN

5th CORPS

LITTLE ROUND TOP

ANDERSON

6th CORPS

TANEYTOWN ROAD

ROUND TOP

OAK RIDGE

McPHERSON RIDGE

CHAMBERSBURG PIKE

MUMMASBURG ROAD

HARRISBURG ROAD

YORK TURNPIKE

Rock

Creek

EMMITSBURG ROAD

Willoughby

Run

⊟ Union cavalry

0 Miles · · · · · 1 · · · · · 2

0 Kilometers · · 1 · · 2

Jeffrey L. Ward

that Confederate artillery was much poorer than Federal guns in calibers and ammunition.[39]

Lee personally gave General George Pickett his orders. He was to cross 1,400 yards of open ground and attack the visible Copse of Trees on Cemetery Ridge about half a mile south of Cemetery Hill. Pettigrew and Trimble were to deploy on Pickett's left, or north, and take up the assault at the same time. Wilcox and Lang were to move on Pickett's right after he had advanced. The attacking force numbered about 13,500.[40]

The attack was to go forward as soon as Confederate artillery, commanded by Porter Alexander, had silenced the enemy batteries. Meanwhile, Stuart's cavalry were to ride to Meade's rear to pursue the Federal army in case the Confederates achieved a breakthrough.[41]

By 10:00 A.M. Alexander had seventy-five cannons disposed around the Peach Orchard and along Seminary Ridge, and another sixty-three guns to his left. Lee did not call upon Ewell's artillery, even though it was located to the north of Cemetery Hill and Ridge and thus could have enfiladed or fired against the flank of the Union guns and infantry. As Alexander wrote, "established where it could enfilade others [it] need not trouble itself about aim. It has only to fire in the right direction and the shot finds something to hurt wherever it falls. No troops, infantry or artillery, can long submit to enfilade fire."[42]

About noon Alexander took a position on Seminary Ridge, where he could observe the effects of the cannonade. Longstreet had ordered him to give the order to Pickett to attack as soon as he saw that the enemy guns had been silenced.

At 1:00 P.M. the Confederate signal guns fired, and in another minute all 138 Rebel cannons were belching forth smoke and shell, the fire converging on about a quarter of a mile of the Federal line along Cemetery Ridge, from Ziegler's Grove south to the Copse of Trees. The enemy guns returned the fire, and a fantastic roar spread over the battlefield. The Confederate guns did much damage to the ridgeline and to Meade's headquarters on Taneytown Road just behind the ridge.

Alexander had planned only a fifteen-minute barrage, but, seeing that the enemy was protected by stone walls and swells of the ground, he couldn't bring himself to give the word to charge. "It seemed madness to launch infantry into that fire, with nearly three-quarters of a mile to go."[43] Alexander wrote Pickett at 1:25 P.M.: "General, if you are to advance at all, you must come at once, or we will not be able to support you as you ought. But the enemy's fire has not slackened materially, and there are still eighteen guns firing."[44]

Five minutes later, Union fire suddenly lessened and the eighteen guns vacated the position. Alexander didn't know it, but Henry J. Hunt, the Union artillery commander, had sent the guns to the rear to cool or to be replaced. Since Alexander had never seen the Federals withdraw guns to save for an infantry fight (something the Confederates did routinely), he concluded that Pickett's advance might succeed if the guns didn't reappear in five minutes. He waited this period, and with no sign of the guns wrote Pickett: "The eighteen guns have been driven off. For God's sake come on quick, or we cannot support you. Ammunition nearly out."

Pickett had already taken Alexander's earlier note to Longstreet and asked, "General, shall I advance?" Longstreet wrote later: "I was convinced that he would be leading his troops to needless slaughter, and did not speak. He repeated the question, and without opening my lips I bowed in answer. In a determined voice Pickett said: 'Sir, I shall lead my division forward.' "[45] Pickett galloped off and put his advance in motion.

Longstreet came out alone to where Alexander was standing. "I don't want to make this attack," he told his artillery chief. "I believe it will fail. I do not see how it can succeed. I would not make it even now, but that General Lee has ordered and expects it."

Alexander believed Longstreet was on the verge of stopping the attack, "and even with slight encouragement he would do it." But Alexander, a colonel, wrote: "I was too conscious of my own youth and inexperience to express any opinion not directly asked. So I remained silent while Longstreet fought his battle out alone and obeyed his orders."[46]

At this moment Pickett's division swept out of the wood and materialized in the full length of its gray ranks and shining bayonets. Joining in on the left, Pettigrew's line followed by Trimble's stretched farther than Alexander could see.

General Dick Garnett, whom Stonewall Jackson had dismissed for or dering a retreat at Kernstown in 1862, passed with his brigade. Alexander rode with him a short distance. They were close friends, having served together in the Great Plains before the war. "We wished each other luck and goodbye, which was our last," Alexander wrote. A few minutes later Garnett was dead.

Alexander rushed back to look for guns still with ammunition to support Pickett's advance. He found fifteen, and they advanced a few hundred yards across the plain, Alexander with them. There they unlimbered and opened fire over the heads of the advancing Rebels.

The Confederate infantry had no sooner appeared on the plain than all of the Union artillery burst out in a thundering barrage. The eighteen guns were back in place. A storm of shell struck the infantry. At the same time all of the Confederate guns on Seminary Ridge opened over the heads of the infantry.

The Federals arrayed along Cemetery Ridge watched the Confederate charge with admiration. Abner Doubleday, observing from his position on the ridge, noted that "the Rebels came on magnificently. As fast as the shot and shell tore through their lines, they closed up the gaps and pressed forward."[47] Lieutenant Colonel Edmund Rice of the 19th Massachusetts, part of Colonel Norman J. Hall's brigade positioned just south of the Copse of Trees, wrote: "From the opposite ridge, three-fourths of a mile away, a line of skirmishers sprang lightly forward out of the woods, and with intervals well kept moved rapidly down into the open fields, closely followed by the line of battle, then by another, and yet a third." The men of John Gibbon's division, holding the Federal center, watched with appreciation as the Confederates came forward with easy, swinging steps. Men exclaimed: "Here they come! Here they come! Here comes the infantry!"[48]

On the right of Pickett's advance, James L. Kemper's brigade was moving forward with its flank in the air. It began to suffer heavy losses from enfilade fire from guns on Little Round Top. Pickett sent repeated messages to Wilcox and Lang to come up on Kemper's flank, but they started late, got lost in the gunsmoke, drifted to the south, and never offered any protection to Kemper.

Kemper's line of march, aiming at the Copse of Trees a bit to the north, took the brigade directly across the front of George J. Stannard's Vermont brigade, the most northern element of Doubleday's division, just to the south of Gibbon's forces. Stannard's men poured heavy fire into Kemper's flank. After marching three hundred yards, Kemper changed front to the right, and the men, breaking into a wild Rebel yell, advanced directly on Norman Hall's brigade, slightly south of the copse.

Stannard saw his chance. He threw his brigade forward a hundred or so yards in front of the Union line, turned it right face, and directed deadly fire into Kemper's exposed right flank. When Kemper hurried back for help, Lewis A. Armistead's brigade, on Kemper's left rear, came forward in a fast charge and closed up on Dick Garnett's brigade ahead. At this moment Kemper fell from his horse, critically wounded.

Garnett had been pushing straight ahead, leading his men through a hail of artillery fire straight toward the Copse of Trees and a stone wall fifty yards in front of it. When the brigade was halfway up the slope, the Federal line rose from behind the stone wall, unleashed a devastating blast from polished muskets that witnesses remembered were glistening in the sunlight, and struck down a huge part of the Confederate column.

The Rebel brigade was still not close enough to charge. Garnett rode along the line, steadying his men. "Don't double-quick," he yelled. "Save your wind and ammunition for the final charge." At that moment his horse fell and Garnett dropped, hit in several places. He never rose again.

Stannard's Vermont brigade had virtually destroyed Kemper's brigade, which staggered and advanced no farther. The Vermonters now sent heavy fire into the flanks of Garnett and Armistead, as these two brigades, now joined in a confused mass, rushed at the stone wall.

From the Copse of Trees, Alexander S. Webb's Pennsylvania brigade commenced an irregular, hesitating fire, while shellfire tore gaps in the Virginians' lines. Canister and grape whirred over the Rebels, sounding to one man like a flock of quail rising in sudden flight. Norman Hall's brigade held its fire, waiting for the Rebel line, whose right was approaching its own right.

The ground dipped and the Rebels were lost to view. "An instant later," Lieutenant Colonel Rice wrote, "they seemed to rise out of the earth, and so near that the expression on their faces was distinctly seen. Now our men knew that the time had come." Aiming low, they opened a deadly discharge upon the mass. Nothing could withstand this merciless fire. The line staggered, hesitated, answered with some wild firing that increased to a crashing roar, and then seemed to melt and drift away.

Surviving Rebels rushed at the stone wall in Webb's sector, using their muskets like clubs and stabbing with their bayonets. No one had time to reload. Canister-shotted cannons fired their last rounds into the faces of the Virginians before the gunners ran away.

At the front of the Rebel charge was General Armistead, holding his black hat aloft on his sword point for the brigade to see. All at once he was over the wall, and 150 of his men rushed behind him. Some of Webb's Pennsylvanians ran pell-mell back over the ridgeline. Armistead put his hand on a Federal cannon. At that instant, he fell mortally wounded. Half of those who crossed the wall with him were killed as well.

The Virginians' triumph was short-lived. No one came to their support, as Pettigrew's and Trimble's men were still pressing the Federals to the north. The Virginians now faced Webb, who remained at the front with the 72nd Pennsylvania, very near where Armistead fell. He had posted a line of wounded men in the rear to drive back or shoot every man who deserted his duty. A portion of the 71st Pennsylvania, behind a stone wall to the north, threw in a deadly flanking fire, while the remainder of the 71st and the 69th Pennsylvania held firmly at the Copse of Trees. As General Doubleday wrote, this was "where our men were shot with the Rebel muskets touching their breasts."[49] At this critical moment Colonel Hall led a splendid charge with two regiments, the 19th Massachusetts and the 42nd New York, passing completely through Webb's line and engaging the Confederates in hand-to-hand combat around their battle flags at the edge of the copse.

As the Union regiments passed the Rebel line that had formed along the stone wall, Rice "could see the men prone in their places, unshaken, and firing steadily to their front, beating back the enemy. I saw one leader try several times to jump his horse over our line. He was shot by some of the men near me." The two Union regiments were driving forward almost at right angles to the remainder of the brigade, and men were falling fast from Rebel fire.

"I could feel the touch of the men to my right and left, as we neared the edge of the copse," Rice wrote. "The grove was fairly jammed with Pickett's men, in all positions, lying and kneeling. Back from the edge were many standing and firing over those in front. By the side of several who were firing, lying down or kneeling, were others with their hands up, in token of surrender."

A Confederate battery firing from near the Peach Orchard tore huge holes through the Federals pushing toward the copse. To get past this line of cannon fire, Lieutenant Colonel Rice saw that his men had to rush forward only a few yards: "I motioned to advance. They surged forward, and just then, as I was stepping backward with my face to the men, urging them on, I felt a sharp blow as a shot struck me, then another; I whirled round, my sword torn from my hand by a bullet or shell splinter. My visor saved my face, but the shock stunned me. As I went down our men rushed forward past me, capturing battle flags and taking prisoners."

This charge was the last straw for the Confederates. With virtually all their field officers gone, the remnants of Pickett's division fled to the bottom of the hill, then turned back in defeat toward Seminary Ridge. Wilcox and Lang's brigades approached but were driven back.

Critical fighting now moved to the Confederate left, as Pettigrew and Trimble assaulted just to the north of Pickett, attacking a stone wall that was being defended by Alexander Hays's Federal division.[50] The Confederates' immediate objective was an orchard just east of the ridge crest. The ground offered no protection here, with the Rebels being particularly exposed to oblique fire from batteries in the cemetery and in Ziegler's Grove. The Confederates took fearful casualties as they neared the crest, but they pressed onward. They did reach the orchard, though by the time they got there they were alone and much diminished in numbers. Pickett's line of battle had disappeared. There was nothing to do but retreat. The troops who advanced the farthest were Captain E. Fletcher Satterfield's company of the 55th North Carolina. Here, at the high-water mark of the Confederacy, Satterfield and other North Carolinians died.

As soon as Porter Alexander saw that the charge had failed, he stopped firing, though he held his ground boldly, afraid Meade would launch an immediate counterattack. But Meade, grateful that the charge had failed, was in no mood for aggression.

As the survivors walked or hobbled back, Lee came forward, alone, to Alexander's advance position and remained for a long while. He was joined by the visiting British Colonel Arthur Fremantle. Lee spoke to the wounded as they passed. General Wilcox came up, almost crying about the state of his brigade. "Never mind, General," Lee said, "all this has been my fault—it is I that have lost this fight, and you must help me out of it in the best way you can."[51]

On the night of July 4, as Longstreet and Lee were standing around a little fire, Lee said again that the defeat was all his fault. But at another time he told Longstreet: "You ought not to have made that last attack." Longstreet replied: "I had my orders, and they were of such a nature there was no escape from them."[52]

The three-day battle of Gettysburg was the bloodiest in American history—50,000 Americans killed, wounded, captured, or missing. The Confederates suffered irreparable losses, 27,000 men—more than a third of Lee's army. In Pickett's Charge alone, about half of the 13,500 Confederates who made the attack were killed, wounded, or captured. The survivors of Pickett's division actually abandoned the battlefield, a sight never before witnessed in the Army of Northern Virginia.

After Pickett's disastrous repulse, there would be no improvement in Confederate strategy and tactics. And though Lee asked a subordinate general to "help me out of" the hole he had dug for the Confederates, in fact there would be no more mystifying maneuvers to rescue the Rebels like the ones Stonewall Jackson had carried out in the Shenandoah Valley and at Chancellorsville. Pickett's Charge was not only the high-water mark of the Confederacy; indeed, it marked the moment when the South lost the war.

＊   ＊   ＊

ON JULY 4, 1863, while the two armies at Gettysburg were eyeing each other across the devastated battlefield, General Pemberton surrendered

Vicksburg and all 37,000 men of its garrison. This was just the latest in a series of disasters that had struck the Confederacy in the West after Braxton Bragg withdrew disgracefully from Kentucky following the battle of Perryville on October 8, 1862.

Even after pushing back Union General William S. Rosecrans and the Army of the Cumberland at the battle of Stones River on December 31, 1862, Bragg had withdrawn. This allowed Rosecrans to set up in Murfreesboro, a key point on the road to the Army of the Cumberland's main objective—Chattanooga. It would be six months before Rosecrans finally succumbed to intense pressure from Ulysses S. Grant at Vicksburg and General Halleck in Washington to move forward, but when he did, in June 1863, he outflanked Bragg and forced the Confederate general to withdraw all the way to Chattanooga.[53]

Bragg withdrew farther in September, after Rosecrans moved on his rear and forced him to evacuate Chattanooga.[54] Bragg retreated across Tennessee's southern border into Georgia, permitting the Union to seize Chattanooga and advance twelve miles south to Gordon's Mills on Chickamauga Creek. On September 19, Bragg sent a Confederate corps across Chickamauga Creek to try to crush Rosecrans's left flank and ultimately cut the Union army off from Chattanooga. The fighting at Chickamauga was intense and bloody—the South suffered 18,000 casualties, the North 16,000. Only the brilliant, dogged defense by Union General George H. Thomas (who would henceforth be known as "the Rock of Chickamauga") kept the Confederates from routing the entire Federal army. Thomas permitted the Union army to re-form and to withdraw unmolested into Chattanooga.[55]

Bragg commenced a siege of the city instead of swinging around it and forcing Rosecrans to evacuate. Rosecrans had an inadequate supply line but did virtually nothing to improve it. After watching for nearly a month as Rosecrans simply sat at Chattanooga while his army began to starve, President Lincoln ousted the general and named Ulysses S. Grant as commander in the West. Grant opened a much shorter supply line and called up the old Army of the Tennessee, now under William Tecumseh Sherman, to raise the siege and secure Chattanooga. Between November 21 and 25, 1863, Grant, Sherman, and Thomas seized Lookout

Mountain and Missionary Ridge, three and a half miles east of Chattanooga, and sent Bragg's army reeling. This latest disaster at last convinced President Davis to remove Bragg, who was replaced by Joseph E. Johnston. More important, the battle of Chattanooga gave the Federals the opening they needed to move into north Georgia and strike at Atlanta.

As the year 1863 ended, then, the Confederacy faced dismal prospects. In the West, the Federal army increased in numbers and built up vast supply and arms caches at Chattanooga with the intention of striking for Atlanta in the spring. In the East, Lee viewed a parallel increase in Federal troops, arms, and supplies. Though General Meade had made no significant aggressive move after Gettysburg, the North's buildup clearly indicated that the Army of the Potomac intended to have a showdown with the Army of Northern Virginia come spring as well.

Appomattox

AS THE YEAR 1864 opened, there was only one residual hope for Southern success: that the people of the North might give up the struggle because the war had become so costly in lives and such a tremendous burden. There was no possibility that Abraham Lincoln and the radical Republicans in command of the Congress would give in. But the Democrats were emerging as a peace party, with a potential peace candidate—General George B. McClellan. If the South could fend off Northern armies until the presidential election in November 1864—if Joseph E. Johnston could keep Federal forces from advancing down the single-track railroad from Chattanooga to Atlanta, and if Robert E. Lee could prevent the Union army from seizing Richmond—the Democrats might take power and negotiate a peace with the South.

There was indeed great despondency in the North. After two and a half years of intensive war, the Union still faced immense barriers to defeat the South. Though the trans-Mississippi states of the Confederacy were isolated and only parts of Mississippi and Alabama south of the Tennessee River remained unoccupied, the South still retained a bastion of great strength—the four eastern states where Southern culture had originated: Georgia, North and South Carolina, and Virginia. President Lincoln, recognizing that a wave of pacifism was sweeping the North, knew that the only way to quell this negative force was for Northern armies to achieve decisive victories prior to the election.

In the East, Union armies were little farther along than they had been when the war started in 1861. Though Gettysburg had destroyed the last offensive power of the Confederacy, Lee's depleted Army of Northern Virginia still stood defiantly in the path of the Federal forces seeking to capture Richmond.

In the West the prospects were brighter for the Union. There Confederate commanders had been consistently inferior to Union leaders. The North had seized Chattanooga and was poised to drive to Atlanta, the lower anchor of the Southern heartland, pivot of the remaining rail links between East and West, and site of vital foundries, machine shops, and munitions factories. But to get to Atlanta, the Federals had to follow the railroad through the forbidding forests and tortuous mountains of north Georgia.

In the early months of 1864, Lincoln resolved to find a general who would drive against Lee in the East no matter how humiliating the defeats he suffered, and who would press Joe Johnston along the difficult road to Atlanta. On March 4, 1864, Lincoln conferred on Ulysses S. Grant command of all Federal armies, directed him to supervise the campaign in Virginia, and promised a steady supply of men and arms until the South collapsed.

Grant named William Tecumseh Sherman to command in the West, and developed a grand strategy designed to win the war. He designated Lee's army, not Richmond, as the real objective in Virginia, and told Sherman to capture Atlanta.[1]

The tasks of both Lee in Virginia and Johnston in Georgia were the same—to prevent the advance of much larger Northern armies. Grant had 144,000 men, plus 16,000 cavalry, against Lee's army of 62,000 men. Sherman had assembled an army just short of 100,000 men, with almost as many guarding his railway supply line, which extended back from Chattanooga to Louisville, Kentucky; Johnston had about 60,000 men, with a corps under Leonidas Polk on the way from Mississippi.

Throughout the war, the manner in which battles and campaigns were fought had been fairly consistent. Commanders on both sides, with few exceptions, had driven straight ahead into the opposition and tried to shatter resistance by main force. Commanders recognized that this

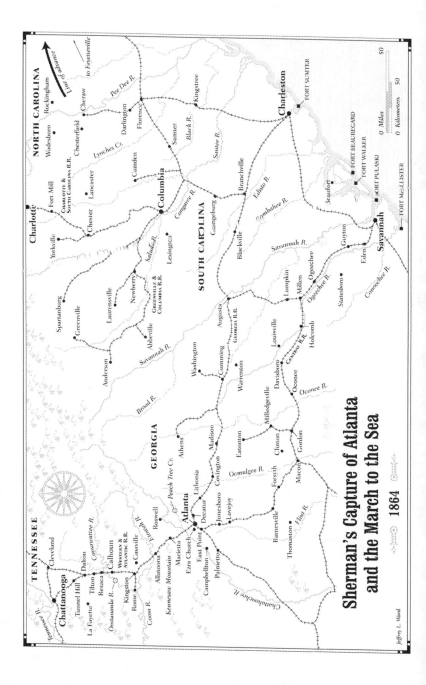

Sherman's Capture of Atlanta
and the March to the Sea

1864

Jeffrey L. Ward

method caused appalling casualties and that it generally didn't work. But no one could think of another tactical method of closing with the enemy.

Few generals had grasped the alternative method of strategic maneuver, which Stonewall Jackson had carried out so perfectly. Maneuver required great imagination and a clear understanding that indirect methods could strike undefended or ill-defended targets, distract and dislodge the enemy, and lead to victory. Most generals on both sides were guileless, uncomplicated warriors who did not understand the wisdom of Paris in the Trojan War, who directed his arrow at the foremost Greek champion Achilles' only vulnerable spot, his heel.

This pattern was about to change, but only in one of the two arenas where the war was going to be decided. And of the four protagonists—Grant, Lee, Sherman, and Johnston—only one, Sherman, adopted the alternative strategy of maneuver. The campaign in Virginia was fought from first to last by the old headlong practices. The other campaign, in Georgia, was won by maneuver. There, Sherman adopted precisely the methods that Stonewall Jackson had used in Virginia. Sherman had shown little evidence of strategic imagination in the battles he had fought previously in the West, and he had not experienced at first hand the effects of Jackson's marches in Virginia. How Sherman came to adopt them is a mystery. He never explained how he reached his epiphany. In any event, maneuver by Sherman became the means by which the North won the Civil War. Had Jackson been heeded, he could have won the war for the South.

* * *

SHERMAN DEMONSTRATED his radical new system in his first encounter with Joe Johnston's army, in early May 1864.

Johnston had set up strong entrenchments and emplaced two infantry corps to cover the low valley above Dalton, thirty miles southeast of Chattanooga, through which the railway ran. He was certain Sherman would attack him directly, since the Federals were dependent on the railway. Johnston was incapable of conducting a war of maneuver. Prior to the start of the campaign, he had rejected a proposal by President Davis to take the offensive into middle Tennessee, in the hope of forcing Sherman to fall back to save Nashville. Johnston intended to stand wholly on the defensive.[2]

But Sherman acted in an entirely unexpected manner. He advanced 74,000 men to demonstrate against the Dalton defenses, but not to attack them, and sent the 24,000-man Army of the Tennessee through the mountains around Johnston's western flank to attack Resaca on the railroad, fifteen miles south of Dalton. This forced Johnston to give up his Dalton defenses and retreat to Resaca, where once again he set up strong defensive emplacements. But once again Sherman moved around his flank, crossing the Oostanaula River southwest of Resaca and forcing Johnston to fall back twenty miles to Cassville.

The ease by which Sherman evicted the Confederates from Dalton and Resaca indicated that another move around Johnston's flanks was likely to be repeated. Building entrenchments was no solution. But the unimaginative Johnston stuck to the same approach of defending the railroad with fortifications—around which Sherman could continue to move.

Instead of relying entirely on static defense, Johnston could have sent part of his army under cover of the mountains and forests to descend unexpectedly on Sherman's most vulnerable target—the railroad leading back to Chattanooga. This railroad was Sherman's umbilical cord. If the Confederates could break the line repeatedly, even if individual breaks could be repaired fairly quickly, Sherman's army would run out of food and would be forced to retreat. In the rough mountains of north Georgia, there was little food to sustain an army of 100,000.

Sherman himself recognized his vulnerability. He wrote that "railroads are the weakest things in war; a single man with a match can destroy and cut off communications." Wooden bridges and water tanks could be burned, trains attacked from ambush, and tracks torn up. The Confederacy already had a brilliant record of strikes on Union communications. From December 1862 to early January 1863, Nathan Bedford Forrest killed or captured 2,500 Federals, seized 10,000 rifles, burned fifty railroad bridges, and severely damaged the Mobile and Ohio Railroad in a raid into western Tennessee. This ruined an overland advance by General Grant. Early in 1863, 13,000 Rebel cavalry and guerrillas threatened north Mississippi and west Tennessee. To defend against them, the North deployed 51,000 men. In the spring of 1863, Grant's army near Vicksburg numbered 36,000 men, while the forces guarding his lines of communication northward amounted to 62,000.[3]

But Johnston had not learned from the experience of Forrest and other raiders, and never attempted a sustained campaign against Sherman's rail line. He thus was fated to continue to fall back before the Federals, his unchanging method leading to an inexorable withdrawal toward Atlanta.

At Cassville, Johnston hoped to strike an isolated advancing Union column and destroy it. Sherman, however, spread out his army in a net of columns over a wide distance. If any one of the columns encountered a Rebel force, multiple columns could quickly come together to concentrate against this enemy.[4] Rather than ambushing a Union column, Johnston was in danger of being enfolded in Sherman's net. The peril forced Johnston to withdraw south of the Etowah River, deeply demoralizing his soldiers.

In this fashion, Sherman pushed Johnston relentlessly back. Only once, on June 27, 1864, did Sherman attempt a frontal attack. Having been held up by wet weather, he ordered two separate attacks a mile apart against entrenchments on Kennesaw Mountain, twenty-five miles north of Atlanta. Both assaults failed utterly. Sherman immediately called off the attacks and planned another move on Johnston's flank.

Four days later James B. McPherson's Army of the Tennessee slipped around Johnston's western flank and threatened the Confederate rear, forcing Johnston to withdraw once more. By July 8, 1864, Johnston had fallen all the way back to Atlanta. This last retreat was too much for President Davis. On July 17, he gave the command to John Bell Hood, a notorious "fighting soldier," a man of little intellect who had never grasped the change in warfare brought on by the Minié-ball rifle and field fortifications.

Hood still thought the recipe for victory was a headlong attack. Playing directly into Sherman's hands, he crashed into the Federal army along Peach Tree Creek on July 20, lost 4,800 men to the Federals' 1,700, and was forced to retreat. On July 22, he tried again, this time losing 8,500 men to the Federals' 3,700. On July 28, he attacked at Ezra Church, a couple of miles west of Atlanta. There he lost 4,600 men to the Federals' 700. This third terrible defeat undermined the morale of the Rebel soldiers. Hood was well on the way to destroying his army.[5]

On August 31, Sherman cut the last railroads out of Atlanta to the southwest and south. Hood realized he no longer could hold the city, and on September 1, 1864, he evacuated it. On September 2, Sherman telegraphed Washington: "So Atlanta is ours and fairly won." The news electrified the Union, revived hope of victory, and made Lincoln's reelection a certainty.

❋   ❋   ❋

IN VIRGINIA an entirely different sort of campaign materialized because Ulysses S. Grant did not possess the imagination of Sherman and decided to attack Lee's army frontally. He moved into the Wilderness, the chaotic, largely roadless region of second-growth timber, briars, and underbrush where the battle of Chancellorsville had been fought the year before. Grant's aim was to pass Lee's left or western flank, get through the Wilderness, and place his army between Lee and Richmond. But his troops were still deep in the Wilderness when they bivouacked on the night of May 4, 1864.

On the morning of May 5, Richard S. Ewell's Confederate corps collided with Governeur K. Warren's 5th Corps and half of John Sedgwick's 6th Corps two miles west of Wilderness Tavern on the Orange Turnpike. The Federals tried to overwhelm Ewell's forces, but they built entrenchments and held firmly.[6] Meanwhile, A. P. Hill's corps marched east on the Orange Plank Road, two miles south, trying to gain the north-south Brock Road. At Brock Road he collided with Winfield S. Hancock's 2nd Corps and a division of Sedgwick's corps. Grant resolved to destroy Hill's corps before James Longstreet's corps could come up, and ordered one headlong assault after another. All were stopped by Rebels behind rudimentary field fortifications. One Confederate described the fighting as "butchery, pure and simple." At the end of the day Hill's troops were exhausted, their ammunition low, and their lines disarranged and disconnected by the fighting.

Grant, frustrated at the failure of the headlong attacks, ordered them to resume on the morning of May 6 against both Confederate corps. Ewell held solidly against six frontal assaults by Sedgwick's corps. But Hill could not hold, and Longstreet was still some distance away when Hancock launched his attack at 5:00 A.M. The Rebels withdrew westward in fair order, leaving only a twelve-gun artillery battalion under William T. Poague to stop Hancock's entire corps. The guns held valiantly for some time but were about to be overrun when a brigade of Longstreet's corps arrived and drove the Federals back. By 8:00 A.M. both sides had dropped down in exhaustion.

Lee organized a counterattack of three brigades along an unfinished railroad line half a mile south of the Plank Road. This threw Hancock's entire corps into chaotic retreat back to the Brock Road, where the troops began industriously building entrenchments. Longstreet led an attack against the Brock Road, but he was accidentally wounded by his own troops and the attack collapsed.

A situation similar to the third day at Gettysburg came about. Although he knew that Hancock's men were entrenching furiously, Lee ordered a frontal assault against the Union fortifications. The attack lasted half an hour and was of unprecedented ferocity. But it failed and the Confederates suffered enormous losses, while the Federals behind their breastworks were little hurt.

The battle of the Wilderness at last ended. The Confederates had lost 7,600 men, the Federals 17,600. Although beaten, Grant refused to retreat, and slipped his army off eastward to Spotsylvania Court House. Lee barely beat him to this crossroads. Once again Grant launched one frontal assault after another. The carnage at Spotsylvania was even worse than in the Wilderness. Seven Union brigades attacked a salient—the Bloody Angle—in the Confederate line on May 10. The assault failed. On May 12, four Union divisions attacked again and broke the line in hand-to-hand fighting that raged from dawn to 10:00 P.M. The Confederates finally drove back the Federals.

Union losses at Spotsylvania were 18,400, Confederate about 9,000. Again Grant refused to retreat, and swung southeastward again. Lee remained ahead of him. The two armies collided near Hanover Junction on the North Anna River between May 23 and 26. Once more Lee foiled

Grant's effort to break through, and once more Grant swung southeast-ward, coming to rest at Cold Harbor, on the ground of the 1862 battle of Gaines Mill. The Union army had not gotten around Lee, but Grant had inexorably advanced and was only nine miles from Richmond.

On June 1 Grant ordered another frontal assault against the Confederate army, using two corps. The Union soldiers advanced straight into the Rebel line. The fire was so intense that the Federals, except in one place, came nowhere near the Confederates, standing in a single rank behind their hastily dug breastworks. The one breakthrough quickly collapsed. In this single, brief charge the Federals lost 2,650 men, the Rebels only a few.

The 1864 campaign had educated soldiers as to what the war had come down to. Attack into the defensive trinity that had been perfected—infantry rifles, artillery canister, and men secure behind field fortifications—was almost a guarantee of death or maiming. But Grant either was not convinced or didn't care. He ordered another assault on June 3. The universal verdict of Confederate officers was that Grant was no strategist and that he relied almost entirely on the brute force of numbers for success.

On the night of June 2–3, many Union soldiers, knowing what lay in store for them, pinned signs on their backs giving their names and hometowns, so their families could be notified after they had died.

The Union soldiers had only about a hundred yards to cover between their positions and the Confederate trenches. The assault commenced with a vast cheering as the Federal lines of battle rushed forward. They were met by sheets of rifle fire and by cannons pushed up to the firing lines, unleashing double-shotted canister directly into the Federal host. The roar of the firing overwhelmed the senses of everyone on the battlefield, and the noise reached Richmond, where people came out on the streets to listen. Both armies were colliding with their whole strength, more fully than ever before. Within seconds, thousands of Union soldiers fell, dead or wounded. Whole ranks of Federal brigades vanished.

The Union army lost 7,300 soldiers in that horrible, brief charge. Confederate casualties were small, only a few hundred.

Grant demanded more attacks, but, as William Swinton, a Northern historian, wrote, every man believed that further effort was hopeless. When General George G. Meade sent orders to renew the assaults, "no man stirred, and the immobile lines pronounced a verdict, silent, yet emphatic, against further slaughter."[7]

❈   ❈   ❈

GRANT'S DIRECT-ATTACK PLAN had failed.[8] He was being forced to see the futility of frontal assaults and to come up with an entirely new plan. He decided to cross the James River ten miles below City Point (now Hopewell) with most of his army, swing around a Federal force under Benjamin F. "Beast" Butler that Confederate General Pierre Beauregard had bottled up in a corner of the river at Bermuda Hundred, seize Petersburg, and cut Richmond's main railroad connection with the rest of the Confederacy.

Grant's envelopment took place over three days, June 15–17, 1864. During this period Lee refused to believe it and was slow in moving to meet the threat. Grant was just stopped from capturing Petersburg by the brilliant defensive work of Beauregard, who had only a few troops.

Thus the major strategic gain that Grant achieved in the campaign of 1864 was by maneuver, not battle. And at Petersburg he once again was stopped by Confederate troops. The officers and men of the Army of the Potomac felt little elation. The army had lost 62,000 men from the Wilderness to Petersburg. Demoralized and utterly exhausted, as were their Rebel counterparts, the Union soldiers began building entrenchments. The investment of Petersburg began. The lines established remained substantially in the same position to the last days of the war.

Although Lee had blocked Grant's direct strike at Richmond, the move on Petersburg had strategically nullified Lee's tactical gains. Grant had penetrated to the very heart of Lee's stronghold. By getting on Lee's flank and threatening to cut the railways, Grant had placed the Confederates in an impossible position. Lee's army, if it remained at Petersburg, would be slowly eaten away.

Lee could throw the war in Virginia into a stalemate, but only for a while. Defending Petersburg was a strategy of defeat. Grant would not win the war at Petersburg, but Lee could lose it there.

There was only one possibility remaining to the Confederacy. Lee could abandon Petersburg and Richmond, restore a war of movement, march his army southward, pick up various Confederate detachments scattered throughout the South, and present Grant with a daunting task.

The English strategist J. F. C. Fuller argued that Northern forces might have occupied Virginia but would have been unable to move far into central or Piedmont North Carolina because they would depend entirely on a single railroad for their vital supply line. That rail line ran through central North Carolina to Charlotte and beyond.[9] Confederate forces operating on the flanks could have constantly cut the railway, crippling the Northern army.

Grant's difficulty in pursuing the Army of Northern Virginia might have given Lee time to reinforce Johnston and destroy Sherman's army. Even if this had failed, the remaining Confederates could have been concentrated into one army or two cooperating armies that, operating on interior lines, could have exploited the wide expanses of Georgia, the Carolinas, and Virginia, kept contact with Alabama and Mississippi, and possibly surrounded and defeated any enemy columns that penetrated into the region.[10]

But Robert E. Lee was uninterested in embarking on a wide-ranging war of maneuver. What's more, he refused even to consider abandoning Richmond. When an aide, Charles S. Venable, asked him why he didn't give up the city, Lee responded sharply that if he did he would be a traitor to his government.[11] By this refusal to abandon the hopeless position at Petersburg and to turn the war into one of movement, Lee ensured that the Confederacy would die.[12]

❋   ❋   ❋

ALTHOUGH THE NORTHERN public was deeply depressed by the news coming out of Virginia, they were buoyed by Sherman's actions in Georgia.

Sherman had concluded that destroying the Southern people's will to pursue the war was more important than destroying the Confederate army. He believed the Rebel armies would melt away once the people wearied of the war. Sherman felt the only sure solution was to inflict so much damage on Southern property and the Southern way of life that the people would prefer surrender to continued destruction. The Union general, then, had arrived at a view that Stonewall Jackson had articulated for the South back in 1861, when he called for the Confederacy to make "unrelenting war amidst their [the Northern people's] homes" and thus to "force the people of the North to understand what it will cost them to hold the South in the Union at the bayonet's point."

After capturing Atlanta, Sherman ordered the entire population to evacuate, forcing men, women, and children out of the city. Homeless, destitute civilians spread across Georgia, seeking shelter, food, and comfort. Sherman's purpose was to punish every Southerner he could reach for seeking to leave the Union. He wanted to destroy the wealth and if possible ruin the lives of all Southerners in his path. He expanded this program in the months ahead to a vendetta of organized ruination that has no parallel in modern history.

Sherman decided his next move was to march to the sea, living off the country and destroying everything in his path. His target would be Savannah, 220 miles away, or Charleston, 260 miles distant. At either point, Union ships could resupply his army. At Atlanta he was 450 miles from his real base of supplies, Louisville, and dependent on a single railroad that could be cut by Confederate raiders almost anywhere.

General Hood had no inkling of Sherman's idea, turned his back on Sherman's army, and planned to march into Tennessee, seize Nashville, sever Federal rail communications with Louisville, and force Sherman to abandon Georgia by cutting off his supplies. Accordingly, Hood embarked on a march toward Nashville by way of Gadsden, Alabama. Sherman pursued him for a distance but called off the effort, sent General George H. Thomas back to defend Nashville with 71,000 men, and left himself with four corps and a cavalry division under Judson Kilpatrick, in all 60,000 men.

Sherman sent all his sick and wounded back into Tennessee, brought forward supplies for his march, then destroyed the railroad as far back as Allatoona, Georgia, and the bridge over the Etowah River. He divided his army into two wings, each of two corps, the right under Oliver O. Howard, the left under Henry Warner Slocum, while Kilpatrick's cavalry were directly under Sherman. Each corps, its transportation pared to the bone, was to move on a separate road. The army carried 200 rounds of ammunition per man and per cannon and twenty days of rations. But the rations were only for emergencies, for Sherman told his troops to live off the land.

General William J. Hardee was in command of the few Confederate troops in front of Sherman's march. The only immediate force was 7,000 men, composed mainly of Joseph Wheeler's cavalry, who had returned from a raid into Tennessee, and Georgia militia with little fighting potential. About 12,000 more men were in garrisons in various locations.

Just prior to the start of the march, on November 15, 1864, the Federals burned the business part of Atlanta, then abandoned the city.

The march confused the Confederates greatly. Four columns moved on widely separated routes, sometimes fifty miles apart. They could not determine Sherman's actual target. The right wing might be aimed at Macon, the left at Augusta. Instead, Sherman drove between the two straight to Savannah.

A path of desolation 200 miles long and as much as 60 miles wide was cut through the center of Georgia. Houses, barns, and other buildings were burned, crops eaten or destroyed, cattle and horses seized, the people reduced to destitution. At Savannah, Hardee had gathered 15,000 troops, but his orders were to abandon the city rather than sacrifice the troops. He withdrew on December 20, 1864.

On reaching the Atlantic, Sherman opened communications with the Union navy. He received a bizarre message from Grant telling him to form a base on the coast with his artillery and cavalry and send the bulk of his infantry to Virginia to help in the campaign against Lee. This astonishing order demonstrated Grant's lack of strategic insight. Sherman's army would be far more decisive if it advanced on Lee's rear through the Carolinas than if it was brought around to attack him frontally. Sherman was able finally to get Grant to rescind the order and give him permission to march northward through the Carolinas.[13]

While Sherman was blazing his destructive path to the sea, Confederate General Hood had come upon a strongly entrenched Union force of about 13,000 men under General John M. Schofield at Franklin, twenty miles south of Nashville. On November 30 Hood threw his men into repeated frontal assaults against the Federal emplacements. The attacks failed utterly and cost him 4,500 men, triple the losses of the Northern defenders. This was the final blow to Confederate morale, for the men realized Hood's tactics were destroying them.

Schofield withdrew to Nashville, where Thomas's entire army was concentrated. It was folly for Hood to follow, but he did. Thomas struck on December 15, throwing the bulk of his army against the left flank of the entrenched Confederates, forcing Hood to a shorter line two miles south. Despite the fact that Thomas had launched a frontal attack, his losses were only 1,000 men, demonstrating that the Rebels no longer were fighting with resolve. The next day a sudden Federal infantry attack on a weakened part of the Confederate line caused the entire Rebel army to collapse. About 4,500 Confederates were taken prisoner, but the bulk got away to the south, halting at Tupelo, Mississippi.

Thus by the end of 1864 the Confederate army had fallen apart in the West, Atlanta had been destroyed, and William Tecumseh Sherman had devastated the South with his march to the sea. And now Sherman was on his way north to finally end the Civil War.

❉   ❉   ❉

ON FEBRUARY 1, 1865, Sherman turned north into South Carolina, aiming one wing at Charleston, the other at Augusta. Once more the Confederate defenders were unable to decide where to concentrate their forces, and once more they divided them between both cities. Sherman struck in between, seizing Columbia, the South Carolina capital, on February 16.

The march on Columbia prevented the two halves of the Confederate forces from uniting, severed the main railroads to Charleston, and compelled the Confederates to abandon the port city on February 15. The senior Rebel commander, Pierre Beauregard, ordered his scattered forces to assemble at Chester, forty-five miles north of Columbia, to protect Charlotte, North Carolina, and the railroads leading to Richmond.

But Sherman avoided Beauregard's force, sending his army northeast in numerous wide-spreading columns through Cheraw, South Carolina, to Fayetteville, North Carolina. The Federals would continue on northeastward to Goldsboro to meet up with Schofield, whose corps of 21,000 men had been sent there by way of New Bern, a nearby North Carolina port that the Federals had held since 1861. At Goldsboro, Sherman expected to resupply his army.

In the crisis, President Davis reinstated Joseph E. Johnston as commander in the Carolinas. Johnston realized that the only possible way to stop Sherman was to exploit the central position between Grant's army at Petersburg and Sherman's army in eastern North Carolina. To do so, Lee had to bring down a substantial portion of his army from Virginia to unite with Johnston's 40,000 men.[14] A now superior Confederate army could defeat Sherman, then turn back on Grant. Sherman had feared this strategy from the start of the Carolinas campaign, questioning whether Lee "would permit us, almost unopposed, to pass through the states of South and North Carolina, cutting off and consuming the very supplies on which he depended to feed his army," and remarking that "if Lee is a soldier of genius, he will seek to transfer his army from Richmond to Raleigh or Columbia; if he is a man simply of detail, he will remain where he is, and his speedy defeat is sure."[15]

Johnston proposed such a strategy to Lee on March 1, 1865, but Lee replied that he was unwilling to turn against Sherman until the Federals had reached the Roanoke River, only fifty-five miles south of Petersburg. This demonstrated Lee's lack of strategic vision and eliminated any possibility of defeating Sherman.

On March 15 Sherman crossed the Cape Fear River at Fayetteville, feinting north with his left wing toward Raleigh to make Johnston and Hardee, the commander on the spot, believe that the North Carolina capital was his target, when in fact it was Goldsboro.

Lee wired Johnston on March 14 that unless he could strike a blow against Sherman, Lee's army would be forced to evacuate Petersburg for lack of supplies. This spurred Johnston to seek battle. He sought to catch one part of Sherman's army isolated and out of reach of the rest. Getting word from his cavalry on March 17 that Goldsboro was Sherman's objective, Johnston set his army in motion for Bentonville, ten miles west of Goldsboro, where he hoped to intercept the left wing of the Union army.

Sherman, thinking Johnston's efforts were aimed not at battle but at protecting Raleigh, ordered his right wing to turn toward Goldsboro, thereby leaving the two lead divisions of his left wing, under General Jefferson C. Davis, alone when they reached Bentonville at midday on March 19. There they bumped into a long line of Confederate entrenchments. The Union troops tried to carry these emplacements by a quick charge but found them too strong, and backed up and dug in. The Confederate right arm under Hardee enfolded the Federal line, rolling back Davis's left flank, but Braxton Bragg—the once supreme Western commander now leading the left arm—called for reinforcements. Johnston sent over a division that otherwise might have swept behind Davis's left and caused disaster. This gave the Federals just enough time for the two other divisions of Sherman's left wing to come up and form a solid line. By nightfall, the Confederate attack had clearly failed, and Johnston pulled his army away.

Sherman continued on to Goldsboro, completing a march through enemy territory, 425 miles, second in length only to Alexander the Great's march through the Persian empire (334–331 B.C.). Sherman's march had driven a stake into the heart of the Confederacy.[16]

❊   ❊   ❊

THE DENOUEMENT NOW came quickly, for Lee's army was ready to collapse at the first heavy blow. The Confederate army had disintegrated partly as a result of the strain of trench life and partly as a result of hunger, as Sherman's advance had contracted supply sources. But the greatest reason for the disintegration was letters from home that reflected the despair of families who had watched Sherman make unchecked progress and destroy their property. Soldiers turned to their fundamental loyalty, their families, and deserted in great numbers to get home to protect those dearest to them.

Though Lee had refused to turn against Sherman, he did in fact recognize that Sherman's march was destroying his own army. He had written the governor of North Carolina on February 24: "The state of despondency that now prevails among our people is producing a bad effect upon the troops. Desertions are becoming very frequent and there is good reason to believe that they are occasioned to a considerable extent by letters written to soldiers by their friends at home . . . that the cause is hopeless and that they had better provide for themselves."[17]

On March 26, Lee admitted to President Davis that he could not prevent the junction of Grant and Sherman and must move his army out of the way. It was much too late, and even now he was reluctant to go.[18] Grant forced his hand on March 28, moving around Lee's extreme right flank, below Petersburg, with the aim of cutting the last two rail lines serving Lee—to Lynchburg and Danville. At Five Forks, north of Dinwiddie Court House, Federal forces routed an unwary force under George Pickett on April 1, sweeping away the last element keeping Grant from driving entirely around Lee's army. The next morning Federals assaulted the whole Petersburg line. In places they broke through. Lee wired Davis that he was withdrawing. The retreat began that night and ended seven days later at Appomattox, where Lee surrendered to Grant in the house of Major Wilmer McLean, who had moved from the farm he owned on Bull Run and whose house had been General Beauregard's headquarters on July 18, 1861, just before First Manassas.

When Lee returned to Confederate lines after the surrender, a number of men were waiting in formation for him. They started to cheer, but something about Lee's somber countenance stopped them. Without a word, they broke ranks and rushed toward him. "General, are we surrendered?" The question was on a thousand lips.

Lee took off his hat. Emotion welled up in him. He tried to hide it, but he could not. At last he was able to speak: "Men, we have fought the war together, and I have done the best I could for you. You will all be paroled and go to your homes until exchanged." Tears came into his eyes. "Be as good citizens," he told them, "as you have been good soldiers." He tried to say more but could bring forth nothing more than a choking "Goodbye" as he rode on.

Lee's admonition to his troops to be as good citizens as they had been good soldiers was of tremendous importance in reconciling the Southern people. By turning their thoughts from continuing the struggle to a peaceful, united future, Lee changed the course of history.

For the remainder of his life Lee devoted himself to the reconciliation of the two sections of the country. He rejected many offers by industrialists to become a wealthy executive. Instead, he took a low-paying job as president of Washington College, in Lexington, Virginia, now Washington and Lee University. There he commenced a deliberate program of educating a new generation of Southerners to be patriotic, loyal citizens of the united nation. Lee's example did more than anything else to end rancor and to reconcile the Southern people.

Lee's failures as a commander prevented the South from gaining independence. But more than any other American, he made it possible for North and South to come back together to create the greatest and most prosperous nation in history.

Notes

Introduction: No Victory Is Inevitable

1. Only one Union commander, William Tecumseh Sherman, was even in Lee's league when it came to overall leadership in battle, and they never met in the field. Lee demonstrated his mastery of battle tactics on many occasions. For example, at Antietam in 1862, though he should not have fought there in the first place, his skillful management of the battle saved the day for the Confederacy at a time when he had fewer than half the troops that Union General George McClellan possessed. He rescued the Confederates by pulling troops from one end of the line and sending them to the other, which was under great stress. Likewise, the Overland Campaign of 1864 stands to this day as one of the greatest defensive operations in history; Lee destroyed half of Ulysses S. Grant's army in the space of weeks and left the Union force completely hobbled for months.

2. Henderson, vol. 1, 175.

3. Clausewitz, 177.

4. Henderson, vol. 1, 201.

Chapter 1: "There Stands Jackson Like a Stone Wall"

1. McDowell, forty-two years old, was a humorless Ohio West Pointer (class of 1838) who had served with distinction in the Mexican War but had been satisfied since with a staff appointment to Winfield Scott, chief of the army. Beauregard had been part of the same military academy class as McDowell and likewise had distinguished himself in the Mexican War. He was briefly superintendent of West Point in 1861 before his Southern sympathies caused his removal.

2. A Virginian, Scott believed the South's love for the Union was as deep as his own, and in time Southerners would see their error and renounce the men who had led them away from where their devotion and true interests lay. Scott said the North should capture New Orleans and seal up the Southern ports, form a large army to move down the Mississippi River, and cut the western Southern states off. Newspapers called it the "Anaconda Plan," likening it to the way an

anaconda slowly chokes its victims. Scott then wanted to hold on tight, neither advancing nor yielding ground. Because of the blockade, Scott said, the South would become a political and economic wilderness. The rebellion would die for lack of resources or be smothered by sheer boredom. Unionist sentiment would reassert itself. The people would force their leaders to sue for readmission to the Union. Scott opposed a massive invasion of Virginia. But Abraham Lincoln and the Republican leadership rejected his passive plan. They wanted to crush the Rebels by military force. Even so, Lincoln followed faithfully Scott's recommendations for a sea blockade and seizure of the Mississippi, establishing a naval blockade of the South on April 19, 1861, and devoting a substantial part (though not the major part) of the North's military efforts to opening the Mississippi River. Scott's naval strategy quickly shut the South off from normal oceanic commerce. By the end of April 1862, New Orleans and all but three of the other eight Southern ports linked by railways with the interior had been occupied. But three ports remained open—Mobile, Charleston, and Wilmington—and they kept the Confederacy alive. In 1862–63 alone 400,000 rifles were imported, while a hundred fast blockade runners operated out of Wilmington on an almost regular schedule with Bermuda and the Bahamas.

3. To get away from the war, McLean moved to Appomattox, Virginia, and bought a house near the county courthouse. By strange coincidence, this was the very house in which Robert E. Lee and Ulysses S. Grant met on April 9, 1865, to arrange the surrender of the Army of Northern Virginia and end the war.

4. These brigades were as follows: South Carolinian Barnard E. Bee, who graduated from West Point one year ahead of Jackson, commanded four regiments from Mississippi, Tennessee, and Alabama. E. Kirby Smith of Florida (later commander of the Trans-Mississippi Department) led a mixture of Virginia, Maryland, and Tennessee units. He was wounded in the battle and his place was taken by Arnold Elzey (West Point 1837), from Maryland. Francis S. Bartow, a Georgia planter and Yale graduate, commanded three Georgia regiments and two Kentucky battalions.

5. On the same day, a small brigade under Theophilus Holmes arrived. It had been guarding Aquia Creek Station, on the lower Potomac River a few miles north of Fredericksburg. Beauregard placed Holmes in reserve behind Richard Ewell on the extreme eastern flank.

6. In the event, Miles didn't even do this minor job adequately. He arrived late on the scene with only a fraction of his force, and made no impression. Part of the reason for Miles's failure was that he was drunk during the battle. McDowell relieved him of his job, but the command failure was McDowell's.

7. The idea came from a military surgeon, A. J. Myer. In 1859 he suggested that messages could be transmitted between distant signal stations elevated over

the intervening terrain by waving a flag or light to the left for a dot and to the right for a dash. Myer thus reduced the four signals of the Morse Code (dot, dash, interval of silence, long dash) to just two signals. Alexander and Myer worked out details of the system in January 1860 with stations located at Fort Hamilton and Sandy Hook, New York. Myer's system was simpler than the semaphore method developed by Claude Chappe in France in 1794, which used a set of arms pivoted on a post. On ships at sea, semaphore signalers communicated with other ships by holding a small flag in each hand. The signaler moved his extended arms to different angles to indicate letters of the alphabet.

8. Porter Alexander, *Memoirs,* 30.

9. Beauregard's plan was for the three most easterly brigades to cross Bull Run, envelop the enemy's exposed left flank, and drive it toward Centreville. As soon as the battle was joined, the brigades to the left, or west, would join in. When Johnston arrived on the field, however, as ranking officer he decided only on a demonstration on the extreme east. But orders to Richard Ewell miscarried and were never delivered, while D. R. Jones crossed and waited two hours. By then it was too late, and his brigade was withdrawn.

10. Bull Run's course is generally east-west. But the stream takes a northwesterly turn a couple of miles below the Stone Bridge. This places both Matthews and Henry House hills northwest of where the main Confederate line was located. However, the soldiers perceived the hills as being on the west, and in this narrative the battlefield is treated as being oriented east-west. Thus the approach from Matthews Hill to Henry House Hill is described as from west to east (not, as the compass shows, from northwest to southeast).

11. Porter Alexander, *Memoirs,* 33, 38.

12. All members of the Henry household except Judith Henry had vacated the premises, but she, aged and bedridden, remained. She was killed in her bed, struck by a cannon shot and several musket balls.

13. Porter Alexander, *Memoirs,* 37. Schenck still believed the Stone Bridge was mined. But he sent axemen, one at a time, across the bridge to cut up and pull away trees that Evans had dropped across the Warrenton Pike to form a crude abatis. This job was completed just in time for Keyes's brigade to retreat across the bridge.

14. Johnson and Buel, vol. 1, 190.

15. Robertson, 466.

16. Davis, 232.

17. Ibid., 234–36. See also Erasmus D. Keyes, *Fifty Years' Observations of Men and Events* (New York: 1884), 434–35.

18. Union losses in the battle of Manassas were 460 killed, 1,124 wounded, and 1,312 captured, for a total of 2,896. Confederate losses were 387 killed, 1,582

wounded, and 13 captured or missing, for a total of 1,982. Thus the South actually lost more in killed and wounded than the North. Figures from Johnson and Buel, vol. 1, 194–95.

19. Porter Alexander, *Memoirs,* 43.

20. Johnson and Buel, vol. 1, 192.

21. Ibid., 193.

22. George B. McClellan, *McClellan's Own Story* (New York: Charles L. Webster and Company, 1887), 66–67. See also Henderson, vol. 1, 155.

23. Joseph Johnston was not an effective offensive commander. He was much more comfortable with passive defensive stands. Beauregard was not a significant factor after the battle of Manassas. Although hailed as a victor in the Southern press, the top leadership recognized that Beauregard's military decisions were erratic and his judgment flawed. His preparations for the battle had been poor, and he had been thoroughly duped by McDowell's surprise strike on his left flank and unprepared to deal with it. Beauregard had avoided disaster only because McDowell had made one mistake after another in the actual conduct of the battle. Beauregard was soon sent off to the West. He failed to produce results there. Later in the war he served with distinction, however, in subordinate roles where he was not called on to make major independent judgments.

Chapter 2: A New Kind of War

1. Actually, there was a fourth way to defeat the North—guerrilla warfare—but it was never seriously entertained by any senior Southern leader. A true guerrilla strategy would have required abandoning the entire South to Union occupation and then raiding Federal supply lines, depots, and isolated units. This is an ancient form of warfare, and the most successful, for it ultimately destroys either the will or the ability of the invader to remain in possession of occupied lands. Guerrilla or "Fabian strategy" prevented the Carthaginian general Hannibal from overcoming Rome in the third century B.C. It was the means by which Scotland retained its independence from England for centuries. It hobbled Napoleon's occupation of Spain from 1808 to 1814, and gave the world the modern term for this kind of war (*guerrilla* means "little war" in Spanish).

Guerrillas nevertheless played a significant, though subordinate, role in the Civil War's western theater, where partisans or "bushwhackers" were common in Union-occupied territory, while orthodox cavalry, led by Nathan Bedford Forrest and other Confederate commanders, carried out spectacularly successful attacks on Union positions or units that closely resembled guerrilla strikes. Guerrillas were less prominent in the East, but John S. Mosby was a brilliant partisan leader in northern Virginia throughout the war, and Robert E. Lee actually considered it in the most critical days before the Seven Days battles in 1862. He told

President Jefferson Davis that if Richmond was conquered, he could fall back into the mountains of Virginia. "And if my soldiers will stand by me I will fight those people for years to come." But Lee and the other Confederate leadership wanted from first to last to fight a conventional war and not resort to a partisan strategy. Just before the surrender at Appomattox on April 9, 1865, the Confederate artillery commander Porter Alexander tried to convince Lee to dismiss his army and have the soldiers travel back to their states and there embark on a guerrilla war. But Lee rejected the idea out of hand. The men, no longer under the control of their officers, would be compelled to rob and steal to live. "They would become mere bands of marauders," Lee said. This was not the sort of war Lee—or any other leaders—wanted to fight.

2. This hope was already being shown to be unrealistic. Immediately after the victory at Manassas, the British foreign secretary, Lord John Russell, refused to meet with Davis's envoy in Britain, William L. Yancey of Alabama, and continued to snub him thereafter. Davis still expected the North to grow tired of the war and quit. But, in light of the North's huge rearmament after Manassas, this expectation was unrealistic. A leader who refuses to look at reality is very likely to bring disaster to his country. Yancey returned to Richmond in December 1861 and told President Davis that the South had no friends in Europe: France would do nothing without Britain, and Britain would never recognize the Confederacy. In November and December 1861, there was brief hope that Britain might intervene after all, because of an almost unbelievable political gaffe by Charles Wilkes, the skipper of a U.S. Navy sloop. On November 8, 1861, he stopped the British mail packet *Trent* sailing from Havana, Cuba, to Southampton, England, in the Bahamas Passage and forcibly removed two Confederate diplomats, James M. Mason of Virginia and John Slidell of Louisiana. Britain prepared for war, sending a military force to Canada. But Prince Albert, Queen Victoria's consort, toned down an ultimatum of the foreign secretary, Lord Russell, and Lincoln, who wanted to fight only "one war at a time," delivered Mason and Slidell to the British on January 1, 1862.

3. From Frederick the Great's *Instructions to His Generals,* written toward the end of his reign.

4. Colonel Jefferson Davis's regiment of Mississippi volunteers was about to be overrun by assaulting Mexicans at the battle of Buena Vista on February 23, 1847. At this moment Bragg came forward with an artillery battery and stopped the enemy. The Mississippi regiment counterattacked, with Bragg in the lead. Davis never forgot the favor. See Bowers, 4–5.

5. The South was divided strategically into two almost completely separate theaters of war by the Appalachian chain of mountains, running more or less north-south from northern Georgia to the Potomac. To keep the Confederacy alive, both theaters had to be defended. The two regions were linked by two main railway lines—one from Richmond through Chattanooga to Memphis, the other

from Charleston through Atlanta and Montgomery to Vicksburg on the Mississippi and from there to Monroe, Louisiana, seventy-five miles west of the river. It was not evident to the leaders on either side until late in the war that the key to defense of the South rested on Chattanooga and Atlanta. The lateral railways ran through these towns and connected both theaters with the Confederacy's open ports and with Richmond. Should Chattanooga and Atlanta be lost and the rail lines cut, then to all intents and purposes the Confederacy would be reduced to the Carolinas and Virginia.

If the Confederacy had operated a major covering force in Virginia and based a large and active army on Chattanooga, the North would have found itself with an almost impossible job. The army at Chattanooga could have advanced into Kentucky when opportunities arose (as they did in 1862), or held eastern Tennessee if attacked strongly. In either event, the lateral railways and the great supply states of Mississippi, Alabama, and Georgia would have been protected. Faced with these barriers, the North would have been unable to defeat the South.

The U.S. Navy's ironclad gunboats precluded the Confederacy from holding the Mississippi by main force. The South would have saved itself much pain and loss if it had recognized this fact and abandoned any attempt to defend the river directly. The correct decision was to defend the river indirectly. This could have been achieved for a long time and perhaps permanently by capturing Louisville, Kentucky, the major Federal base in the West. In the early fall of 1862, Confederate armies invaded Kentucky and had a wide-open path to the virtually undefended city. But General Braxton Bragg turned his forces away from Louisville and abandoned the effort. A better general would have seized Louisville. Then Cincinnati could have been threatened and the Ohio River blocked. With boat traffic on the Ohio stopped, its main base at Louisville eliminated, and Confederate cavalry raids into Indiana and Illinois disrupting rail traffic to St. Louis, the North would have found it extremely difficult to mount a campaign to clear the Mississippi. Capturing Louisville would have fundamentally changed the nature of the war. But for the want of a competent commander in the West, the South failed to do so.

The Confederates were left with a strategic situation that demanded abandoning the Mississippi and the states west of it and concentrating every possible soldier into two well-led armies, one holding the approaches to Chattanooga, the other Virginia. Unfortunately for the South, this was not done, and it suffered grievously. In Virginia the army got insufficient troops but did get splendid leaders, Lee and Jackson. In the West, President Jefferson Davis not only failed to provide enough troops but also kept Bragg in command until he had virtually lost the entire theater.

6. Source: letter from Gustavus W. Smith to G. F. R. Henderson. See Henderson, vol. 1, 173–76.

7. Henderson, vol. 1, 174–76. Henderson's source for Jackson's proposal was a personal letter he received from General Smith.

8. Rifles had been around for centuries and were more accurate and longer-ranged than muskets. But the rifling in the barrel quickly became fouled by products of gunpowder combustion. Minié produced a bullet with a hollow base. When the rifle was fired, the explosion expanded the base to fit snugly against the rifling grooves, scouring the fouling of the previous shot from the grooves. A Minié-ball rifle could fire many rounds before the barrel had to be cleaned. The Minié-ball rifle was coupled with the copper percussion or firing cap containing fulminate of mercury, invented in 1816 by Thomas Shaw of Philadelphia. The percussion cap made the rifle musket serviceable in all weathers and vastly reduced misfires.

9. The smoothbore musket was truly effective only at close range. The Prussian army at the close of the eighteenth century set up a canvas target 100 feet long by 6 feet high, simulating an enemy unit's silhouette. At 225 yards, 25 percent of a unit's balls hit the target; at 150 yards, 40 percent; at 75 yards, 60 percent. But if fire was held to 50 yards, horrible casualties resulted, often 50 percent or more of the men. See David G. Chandler, *The Campaigns of Napoleon* (New York: Macmillan Publishing Co., 1966), 342.

10. It will occur to readers that the answer of Generals Bee and Bartow pointed to the necessity for a total change in infantry tactics and to an end to men marching up shoulder to shoulder on the enemy. The Minié-ball rifle made this no longer a feasible way of winning engagements. The solution had to be revolutionary— eliminating ordered ranks of men and stand-up attacks entirely. This was far too complicated a problem to solve by officers steeped in the traditional method of attack.

Indeed, no one could think of an answer in the years after the Civil War, when magazine repeating rifles firing brass cartridges at ranges of well over 1,000 yards replaced single-shot weapons, not to mention the machine gun, perfected by Hiram Maxim in 1884, and the development of far more powerful explosives and far more accurate fuses for far longer-ranged artillery. In the Boer War of 1899–1902, soldiers still tried to advance in formation, albeit in extended order, not shoulder to shoulder. But this was not the answer, and soldiers went into World War I in 1914 advancing in formation. Killings were so stupendous, however, that a solution had to be found. It came in 1915 from a German captain, Willy Martin Rohr, who developed "infiltration" or "storm troop" tactics. One small group of soldiers held down an enemy trench position or strong point with heavy directed fire by automatic weapons, mortars, and sometimes light cannons, while one or more well-trained teams of eight to twelve "storm troopers," working in conjunction, infiltrated the trench line or sneaked up on the strong point and "rolled it up" with grenades, small-arms fire, and sometimes flamethrowers. Soldiers were in no formation at all, and each man had a particular task. This fire-and-maneuver system overcame enemy guns and fortifications, returned movement to the battlefield, and became the fundamental method of tactical engagements to this day. See Bevin Alexander, *How Wars Are Won* (New York: Crown, 2002), 19, 253, 361 n. 15.

11. Porter Alexander, *Fighting for the Confederacy,* 408–9, describes these fortifications at the battle of Cold Harbor in June 1864, when they were reaching near perfection. He includes a sketch of how parapets were constructed: a trench two feet deep was edged by a mound two and a half feet high of dirt, rocks, and logs. Even artillery fire could not breach such defenses. Most unusual was Union General Henry W. Halleck. In spring 1862, advancing on Corinth, Mississippi, after the battle of Shiloh, he ordered his men to throw up embankments and dig foxholes, or "ditches" as they called them, even for a night's stop. And Union General Jacob D. Cox said "one rifle in the trench is worth five in front of it."

12. Cannons had to fire straight at their targets. Civil War gunners usually had to see what they were aiming at and to have an open field of fire to it. They needed fairly level places to locate, since the guns had no recoil mechanisms and absorbed the shock of recoil by rolling backward. Before being fired again, the guns had to be rolled back "into battery," or the original firing positions.

13. Several European rifled cannons were invented in the 1840s, but William Armstrong, an English engineer, developed the first production model in 1855, followed by French and Prussian types. Captain Robert P. Parrott manufactured rifled cannons at Cold Spring, New York. Parrott guns were employed extensively in the war. The new pieces were mostly high-velocity guns used for direct fire. Rifled guns for the first time in war caused serious casualties among troops far in the rear of the front. Also, rifled cannons sometimes caused losses at long range among soldiers attacking in the open in the massed two-man-deep lines of battle that characterized the Civil War.

14. Wolseley, 116.

15. Johnson and Buel, vol. 2, 297.

16. Sherman actually won the war with a bold strike from Atlanta to Savannah in late 1864 that cut the Confederacy in two. But Grant and Sherman were confronting decidedly inferior Confederate commanders in the West, and their successful surprise strokes came in circumstances where Southern commanders had placed themselves in nearly hopeless strategic situations that Grant and Sherman exploited. Just one example: Union armored gunboats permitted Grant to land south of Vicksburg on the Mississippi in spring 1863. Instead of striking directly for the town, he used great imagination in seizing the rail hub of Jackson to the east, isolating Vicksburg. But this would not have been an irretrievable blow if the Confederate Commander John Clifford Pemberton had joined his army with Joseph E. Johnston's force north of Jackson. Instead Pemberton foolishly withdrew into Vicksburg, where he was easily besieged and forced to surrender in July.

Chapter 3: The Shenandoah Valley Campaign

1. Pinkerton reports were usually carried in *War of the Rebellion* under the assumed name of E. J. Allen.

2. During the past winter, Jackson had tried to recover parts of northwestern Virginia beyond the Allegheny Mountains. Beginning on January 1, 1862, Jackson foiled a projected offensive by Union Major General William S. Rosecrans to seize Winchester and threaten the west, or left, of the main Confederate position at Manassas. Jackson drove his small army through snow and rain, attacked Federal garrisons, temporarily broke the Baltimore and Ohio Railroad along the Potomac River, and seized Romney, thirty-five miles northwest of Winchester. This highly successful campaign threatened Federal rail communications with the Midwest and regained a large part of West Virginia at little cost. Officers and men of Jackson's chief subordinate, Brigadier General W. W. Loring, became angry with the conditions they faced in Romney. Eleven officers complained to the Confederate Secretary of War Judah P. Benjamin. Without investigating, Benjamin ordered Jackson to bring the Romney garrison back to Winchester immediately. Jackson complied but sent in his resignation at once. "With such interference in my command, I cannot expect to be of much service in the field," he wrote. The abrupt response shook General Johnston, who recognized Jackson's ability. By his pleadings and the influence of Virginia governor John Letcher, Jackson withdrew his letter. And no secretary of war tried to second-guess Jackson again. The War Department transferred Loring and part of his troops elsewhere.

3. The terrain of the Shenandoah Valley rises to the south. Thus the Shenandoah River drains northward, and one moving south in the Valley goes up, not down.

4. For a full analysis of the *Monitor* and the *Merrimac,* see Johnson and Buel, vol. 1, 692–750.

5. Garnett later became commander of a brigade in George E. Pickett's division and died leading it in Pickett's charge at Gettysburg July 3, 1863.

6. In an article in the "Battles and Leaders" series in *Century* magazine more than twenty years after the war, Johnston, discussing the April 14 meeting, said nothing about a strike into the North, but devoted his whole argument to the need to bring all forces together and to attack McClellan at Richmond. See Johnson and Buel, vol. 2, 203.

7. *War of the Rebellion,* vol. 12, part 3, 890.

8. George H. Gordon, *From Brook Farm to Cedar Mountain* (Boston: James R. Osgood and Company, 1883), 136.

9. Kyd Douglas's account of his ride from near Rudes Hill to General Ewell's headquarters, described vaguely by Jackson as "somewhere near Culpeper Court House," is one of the most fascinating stories of courage, perseverance, and sheer exhaustion recorded in the war. Douglas had never set foot in that part of Virginia

and had only the foggiest idea of where Culpeper was. His journey, then, was something of an odyssey and a voyage of discovery. When Douglas finally found Ewell at his headquarters at Brandy Station, northeast of Culpeper, and handed him the message from Jackson, he was so fatigued he was staggering and about to collapse. As Douglas describes it, Ewell "caught me, led me to his cot, and laid me there; and then the dear, rough, old soldier made the air blue with orders for brandy and coffee and breakfast—not for himself but for me. My ride was done and nature asserted itself by reaction and exhaustion. In less than twenty hours I had ridden about 105 miles." Ewell sent Douglas back to Culpeper in an ambulance. There he fell into a hotel bed and slept for nearly twenty-four hours. See Douglas, 42–46.

10. *War of the Rebellion,* vol. 12, part 3, 848; Henderson, vol. 1, 281. On April 14, 1862, Jackson wrote Ewell: "I hope that Banks will be deterred from advancing much farther toward Staunton by the apprehension of my returning to New Market and thus getting on his rear."

11. Robert Lewis Dabney (Dabney, 334–35) writes that Banks considered a strike southwest from Harrisonburg on the Warm Springs Turnpike (present-day Virginia Route 42) to Churchville west of Staunton, where Milroy's brigade could have joined him from Monterey (over present-day U.S. Route 250). This would have placed the two Federal forces northwest of Edward Johnson's force at West View. This, Dabney writes, "would have ensured the destruction of the little army of Confederates. The two Federal forces united would then have easily occupied Staunton, and made the Valley untenable for Jackson." But Banks didn't dare to undertake this operation because he feared leaving his rear exposed to Jackson, moving from Conrad's Store. "General Jackson stated in his correspondence that he foresaw the danger of such a maneuver, and calculated the timidity of his opponent as a sufficient defense."

12. *War of the Rebellion,* vol. 12, part 3, 859.

13. Jackson's letter is recorded in *War of the Rebellion,* vol. 12, part 3, 872, and in Allan, *Campaign of Gen. T. J. (Stonewall) Jackson,* 68n.

14. Likewise, Banks had been completely finessed by the move. He did not dare to march on Staunton with Ewell in Elk Run Valley ready to pounce on his flank and rear if he did so. By pushing Turner Ashby's cavalry patrols forward, Jackson had also ensured that Banks would not learn of the movement across the Blue Ridge until it was too late for him to do anything about it. It would not have been impossible, however, for Banks to strike for Staunton when he found that Jackson had vanished from the Valley. It was necessary, therefore, that Jackson's absence from the Valley be short. He achieved this my moving his army swiftly from Mechums River Station to Staunton by rail.

15. In this period of the war, there was still too little appreciation of the range and power of the Minié-ball rifle, and too much foolish bravado. The 12th Georgia,

holding the center of the Confederate line on Sitlington Hill, refused to retire to the reverse of the ridge to escape Federal bullets. Their commander tried to withdraw them, but the men resisted and remained on the crest, where they made easy targets from below. A Georgia youth, asked the next day why they did not seek cover, replied: "We did not come all this way to Virginia to run before Yankees." Accordingly, the 12th Georgia lost 156 men and 19 officers. See Dabney, 349.

16. To make doubly sure that Frémont remained caged in the mountains, Jackson dispatched his topographical engineer, Captain Jedediah Hotchkiss, with a few cavalry to block all of the passes through the Alleghenies to the east by means of felled trees and burned bridges. Jackson also ordered John D. Imboden, recruiting troops in Staunton, to seal off any routes by which Frémont might reach Harrisonburg. Imboden knew of a spot four miles east of Franklin on the road to Harrisonburg (present-day U.S. Route 33). "There was a narrow defile hemmed in on both sides by nearly perpendicular cliffs, over 500 feet high," Imboden wrote. "I sent about fifty men, well armed with long-range guns, to occupy these cliffs, and defend the passage to the last extremity." On May 25 Frémont, under orders from Stanton to cut off Jackson's retreat up the Valley, sent his cavalry to feel out passage to Harrisonburg. "The men I had sent to the cliffs let the head of the column get well into the defile, when, from a position of perfect safety, they poured a deadly volley into the close column." The Federal troopers halted in confusion. "Another volley and the 'rebel yell' from the cliffs turned them back, never to appear again." Frémont abandoned the effort to reach Harrisonburg, went north to Moorefield, and emerged into the Shenandoah Valley far to the north at Strasburg. See Johnson and Buel, vol. 2, 290–91.

17. Wolseley, 94–95.

18. Taylor, 43–44.

19. Ibid., 45.

20. George H. Gordon, *From Brook Farm to Cedar Mountain* (Boston: James R. Osgood and Company, 1883), 191–93, and Henderson, vol. 1, 323–34.

21. Richard Taylor recounts an encounter that occurred during the nighttime march toward Winchester. "At dusk we overtook Jackson, pushing the enemy with his little mounted force, himself in advance of all. I rode with him, and we kept on through the darkness. An officer riding hard, overtook us, who proved to be the chief quartermaster of the army [Major John A. Harman]. He reported the wagon trains far behind, impeded by a bad road in the Luray Valley. 'The ammunition wagons?' sternly. 'All right, sir. They were in advance, and I doubled teams on them and brought them through.' 'Ah!' in a tone of relief. To give countenance to the quartermaster, if such can be given on a dark night, I remarked jocosely, 'Never mind the wagons. There are quantities of stores in Winchester, and the general has invited me to breakfast there tomorrow.' Jackson took this seriously, and reached out to touch me on the arm. Without physical wants himself, he

forgot that others were differently constituted, and paid little heed to the commissariat, but woe to the man who failed to bring up ammunition! In advance his trains were left behind. In retreat, he would fight for a wheelbarrow." See Taylor, 51.

22. *War of the Rebellion,* vol. 11, part 1, 31.

23. *War of the Rebellion,* vol. 12, part 3, 220; Porter Alexander, *Memoirs,* 102.

24. Henderson, vol. 1, 347. The Prussian strategist Karl von Clausewitz (1780–1831) wrote in his classic work, *On War* (book I, chapter 7): "The military machine . . . is basically very simple and therefore easy to manage. . . . But none of its components is of one piece: each part is composed of individuals, every one of whom retains his potential of friction. . . . A battalion is made up of individuals, the least important of whom may chance to delay things or somehow make them go wrong." Clausewitz wrote that in war this tremendous friction is everywhere in contact with chance. It produces effects that cannot be measured. Fog can prevent an enemy being seen, a gun being fired, a report reaching the commander. Rain can prevent a battalion from arriving, make another late, bog horses in mud, or cause other problems. "Action in war," Clausewitz wrote, "is like movement in a resistant element. Just as the simplest and most natural of movements, walking, cannot easily be performed in water, so in war it is difficult for normal efforts to achieve even moderate results." In war, friction is the force that makes difficult what is apparently easy. See Peter Paret, ed., *Makers of Modern Strategy* (Princeton, N.J.: Princeton University Press, 1986), 202–3.

25. Porter Alexander, *Fighting for the Confederacy,* 91.

26. A. R. Boteler, *Southern Historical Society Papers,* vol. 40, 165. See also Dabney, 391; Cooke, 158–59; Freeman, *Lee's Lieutenants,* vol. 1, 414n; and Freeman, *Lee,* vol. 2, 83–84.

27. Jackson had already shown, in his movement to Front Royal, that he was capable of deceiving the enemy to such an extent that he could move on virtually any line he chose, irrespective of Federal forces in the way. This capability was shown in brilliant outline in his proposal to Lee on April 29, 1862, in which he suggested that if he could get 5,000 reinforcements to contain Frémont, he would cross the Blue Ridge to Sperryville and move north, threatening both Front Royal and Warrenton, thereby forcing Banks out of the Valley and holding in place all the forces protecting Washington. See note 13 above and *War of the Rebellion,* vol. 12, part 3, 872, and Allan, *Campaign of Gen. T. J. (Stonewall) Jackson,* 68n.

28. Even if, by any chance, McClellan had been able to bring a superior army against Jackson in the North, the Confederate general had already demonstrated that he could isolate parts of the larger Union army in order to secure victory.

29. On June 2, 1862, Colonel John R. Patton of the 21st Virginia reported to Jackson about a rearguard skirmish in which he expressed regret that his men had killed three Union horsemen who had bravely charged into the regiment.

After hearing the report, Jackson looked hard at Patton and asked: "Colonel, why do you say you saw those Federal soldiers fall with regret?" Patton replied they had exhibited vigor and courage, and a natural sympathy with brave men led to the wish that their lives might have been saved. "No," Jackson replied sternly, "shoot them all. I do not wish them to be brave." See Dabney, 397; Robertson, 424.

30. Jackson had sent his 2,300 Federal prisoners ahead to Waynesboro, where they were loaded on trains and sent to Richmond.

31. Allan, *Campaign of Gen. T. J. (Stonewall) Jackson,* 164 n. 2. McDowell's chief of staff wrote Shields on June 9: "If you are in hot pursuit and about to fall on the enemy . . . the general is not disposed to recall you."

32. Garnet Wolseley (Field Marshal Viscount Wolseley), commander in chief of the British army, wrote in the *North American Review* of August 1889 (vol. 149, 165–66) that Jackson's actions at Cross Keys and Port Republic constituted "an operation which stamped him as a military genius of a very high order." Reprinted in Wolseley, 129.

33. On June 1, 1862, Secretary Stanton ordered the 9,000-man division of George A. McCall in McDowell's corps to move by water to reinforce McClellan. It arrived on June 12–13. But Jackson's victory at Port Republic paralyzed the remainder of McDowell's forces. It resulted, finally, in the withdrawal of James B. Ricketts's and Shields's divisions toward Manassas, beginning June 17, and the retention of Rufus King's division at Fredericksburg. On June 20 McClellan had 105,000 men before Richmond, while Federal forces in the Valley and around Washington totaled more than 60,000. This was the fruit of Jackson's Valley campaign. See ibid., 170, and *War of the Rebellion,* vol. 11, part 2, 490.

34. Freeman, *R. E. Lee,* vol. 2, 84, and Lee, 5–10.

35. *War of the Rebellion,* vol. 12, part 3, 910; Freeman, *R. E. Lee,* vol. 2, 95–96. In a letter on June 6, 1862, Lee alerted Jackson that he was likely to be called to Richmond. On June 11, while informing Jackson that he was sending the brigades of Lawton and Whiting, Lee also directed him to move to Ashland. On June 16, Lee wrote Jackson: "The sooner you unite with this army [at Richmond] the better. . . . In moving your troops you could let it be understood that it was to pursue the enemy in your front. Dispose those to hold the Valley so as to deceive the enemy." See Allan, *Campaign of Gen. T. J. (Stonewall) Jackson,* 165, 168–69.

36. *Southern Historical Society Papers,* vol. 40, 172–73; Cooke, 201.

37. *War of the Rebellion,* vol. 12, part 3, 872; Allan, *Campaign of Gen. T. J. (Stonewall) Jackson,* 68n.

38. Freeman, *R. E. Lee,* vol. 2, 102; *Southern Historical Society Papers,* vol. 40, 173–74.

39. MS in *The Centennial Exhibit of the Duke University Library* (Durham, N.C.: Duke University, 1939), 15–16.

Chapter 4: The Seven Days

1. Lee's cavalry chief, Jeb Stuart, had proved this in a famous ride, commencing on June 12. The purpose of his reconnaissance, with 1,200 troopers, was to find out how far north Porter's defensive line extended and whether White House was still in use. After ascertaining that Porter's line went only a short distance and that White House was in full operation, Stuart realized he had stirred up a hornet's nest of Union cavalry in his rear. He decided that the easiest way to return to Confederate lines was to march all the way around McClellan's army. This he accomplished, with scarcely any losses, coming back up the River Road (present-day Virginia Route 5) alongside the James River on June 15.

2. The Army of the Potomac numbered 105,000 men and 25,000 animals. It required 600 tons of ammunition, food, forage, and medical and other supplies every day. In that period the U.S. Army needed about ten pounds of supplies per man per day. At the time of the campaign, from fifty to sixty days' rations for the entire army, or about 20,000 tons of goods, had been accumulated at White House. McClellan's quartermasters kept 400 transports moving constantly between Alexandria and White House. See Henderson, vol. 2, 30–31.

3. The reason for this delay has never been satisfactorily explained. Lee's biographer Douglas Southall Freeman wrote that Lee didn't know whether McClellan was heading for Fort Monroe or the James, and was unwilling to move until he was sure. (See Freeman, *R. E. Lee,* vol. 2, 163.) But the absence of any troop movements toward Fort Monroe (which would have had to pass by Bottom's Bridge, only a couple miles south of Dispatch Station and where some of Richard S. Ewell's infantry were standing) should have told him that the James was McClellan's destination.

4. Moltke achieved fame soon after the American Civil War. But the issues of concentrating forces on the battlefield had been intensely debated in military circles in Europe and America for years prior to the Civil War. Thus Lee's plans for the attack on Porter were evidently influenced by these debates. Moltke's conclusions, written prior to 1885, were considered by military theorists as representing the most incisive thinking on the matter. They are widely at variance with what Lee ordered. Moltke wrote: "Conditions are more favorable when the fighting forces can be concentrated from different points toward the battlefield on the day of battle, and when operations can be conducted so that a final *short* march from different directions leads to the enemy's front and flank simultaneously." In other words, Moltke held that parts of armies should *approach* from different directions but that they should not press forward into battle until all of the parts were in place, ready, and in direct contact with all the other parts. To Moltke it was unthinkable for forces to move from various distant places directly into battle, because this would make no provision for incalculable factors such as the outcome of minor en-

gagements, the weather, or false information. The slightest hesitation might ruin the combination. Haste was even more to be dreaded. There was always the danger that one wing might attack, or be attacked, while the other was still far distant. See Henderson, vol. 2, 19, and *Moltke on the Art of War, Selected Writings,* edited by Daniel J. Hughes (Novato, Calif.: Presidio Press, 1993), 262–63.

5. Lee's order did not require Branch to inform A. P. Hill and specified only that "as soon as the movements of these columns are discovered" Hill was to cross the Meadow Bridge and descend on Mechanicsville, two miles to the east. A. P. Hill's movement, in turn, was to open up passage of the divisions of Longstreet and D. H. Hill over the Mechanicsville Turnpike bridge just south of the village. See *War of the Rebellion,* vol. 11, part 2, 489–99; Porter Alexander, *Military Memoirs,* 113; Henderson, vol. 2, 13–14.

6. Lee not only approved A. P. Hill's attack but ordered D. H. Hill to send a brigade to assist. See Johnson and Buel, vol. 2, 361; Robert E. Lee Jr., *Recollections and Letters of General Robert E. Lee* (New York: 1904), 415; Marshall, 94; Freeman, *Lee's Lieutenants,* vol. 1, 514; *War of the Rebellion,* vol. 11, part 2, 834–40 (A. P. Hill's report); and Freeman, *R. E. Lee,* vol. 2, 133–35. Roswell S. Ripley reports (*War of the Rebellion,* vol. 11, part 2, 648) that Lee and D. H. Hill ordered him to attack. Freeman remarks that "a turning movement upstream [on Beaver Dam Creek] had been regarded as a sine qua non because the strength of Porter's position was so well known." Thus Lee knew that a frontal attack at Beaver Dam would be extremely costly.

7. D. H. Hill wrote afterward: "The blood shed by Southern troops there was wasted in vain, and worse than vain; for the fight had a most dispiriting effect on our troops. They could have halted at Mechanicsville until Jackson had turned the works on the creek, and all the waste of blood could have been avoided. . . . The crossing of the river by General A. P. Hill before hearing from Jackson precipitated the fight on the first day; and it having begun, it was deemed necessary [by Lee] to keep it up, without waiting for Jackson." See Johnson and Buel, vol. 2, 361.

8. After Jackson arrived at Hundley's Corner, he and his troops heard the rapid discharges of cannons to their south. Jackson had been informed by Lee that passage of the river was to be made along with a heavy cannonade. Since the firing lasted only for a short period, Jackson concluded it was part of Hill's passage of the river and the eviction of the Federals from Mechanicsville. Jackson also knew that no attack had been contemplated on Beaver Dam Creek, and thus he did not suspect that A. P. Hill would impetuously make such a blunder and that Lee would support him. Jackson knew, once he arrived at Hundley's Corner, that Porter's line along Beaver Dam Creek was untenable. The Federals would never stand with a strong force on their flank and menacing their communications. Jackson felt that the Federals would fall back during the night, and the Confederate advance would be carried out the next day in the concentrated formation Lee's

orders had dictated. In any event, nothing could have been gained if Jackson had attacked that night, and much would have been risked. Robert Lewis Dabney, Jackson's chief of staff, explained why: "We heard no signs of combat on Beaver Dam Creek until a little while before sunset. The whole catastrophe took place in a few minutes about that time; and in any case our regiments, who had gone into bivouac, could not have been reassembled, formed up, and moved forward in time to be of any service. A night attack through the dense, pathless, and unknown forest was quite impracticable." (Source: a letter to G. F. R. Henderson from Dabney; see Henderson, vol. 2, 23.)

9. Jean Colin, *The Transformations of War* (London: Hugh Rees, 1912), 279–89.

10. Henderson, vol. 2, 25.

11. Johnson and Buel, vol. 2, 337.

12. Lee thought Porter's dispositions were radically defective, since he was not blocking access to White House and the railroad. Old Cold Harbor was Porter's best location for this purpose. One road led from this crossroads directly east to White House and another arched southeast to join a road running parallel to the river from New Cold Harbor to Dispatch Station and on a couple of miles to Bottom's Bridge and the Williamsburg Road. Lee did not consider the opposite possibility: that Porter was deliberately not defending Old Cold Harbor because he was not protecting White House and the railroad. Rather, Lee continued to expect that Porter would be obliged to retreat to shield the roads leading from Old Cold Harbor. This is shown in Jackson's report: "Hoping that Generals A. P. Hill and Longstreet would soon drive the Federals towards me, I directed General D. H. Hill to move his division to the left of the wood, so as to leave between him and the wood on the right an open space, across which I hoped that the enemy would be driven." See *War of the Rebellion,* vol. 11, part 2, 553.

13. E. McIvor Law, commanding a brigade in W. H. C. Whiting's division, wrote a graphic description of the final assault. "General Whiting ordered that there should be no halt when we reached the slight crest occupied by the few Confederate troops in our front, but that the charge should begin at that point in double-quick time, with trailed arms and without firing. Had these orders not been strictly obeyed the assault would have been a failure. No troops could have stood long under the withering storm of lead and iron that beat into their faces as they became fully exposed to view from the Federal lines. As it was, in the very few moments it took them to pass over the slope and down the hill to the ravine, a thousand men were killed or wounded. Law's brigade advanced in two lines, the 11th Mississippi and the 4th Alabama forming the first line, and the 2nd Mississippi and the 6th North Carolina the second. John Bell Hood's Texas brigade had a similar formation on our left. . . . Passing over the scattering line of Confederates on the ridge in front, the whole division broke into a trot down the slope toward the

Federal works. Men fell like leaves in an autumn wind, the Federal artillery tore gaps in the ranks at every step, the ground in rear of the advancing column was strewn thickly with the dead and wounded; not a gun was fired in reply; there was no confusion, and not a step faltered as the two gray lines swept silently and swiftly on; the pace became more rapid every moment; when the men were within thirty yards of the ravine, and could see the desperate nature of the work in hand, a wild yell answered the roar of Federal musketry, and they rushed for the works. The Confederates were within ten paces of them when the Federals in the front line broke cover, and, leaving their breastworks, swarmed up the hill in their rear, carrying away their second line with them in their rout. We then had our innings. As the blue mass surged up the hill in our front, the Confederate fire was poured into it with terrible effect. . . . Firing as they advanced the Confederates leaped into the ravine, climbed out on the other side, and over the lines of breastworks, reaching the crest of the hill beyond with such rapidity as to capture all of the Federal artillery (fourteen pieces) at that point. We had now reached the high plateau in rear of the center of General Porter's position, his line having been completely cut in two, and thus rendered no longer tenable." See Johnson and Buel, vol. 2, 363.

14. Of course, McClellan's spy chief, Allan Pinkerton, had vastly inflated the size of Lee's army. However, it was McClellan's duty to ascertain the true conditions he faced. Napoléon Bonaparte said that a commander who did not know the size and dispositions of his enemy's army was not worthy of command.

15. Lee's orders for pursuit on June 29 were for Magruder, with 11,500 men, to move down Williamsburg Road and engage the Federals before they reached White Oak Bridge; for Benjamin Huger, with 9,000 men, to march eastward on the Charles City Road, south of White Oak Swamp, and take the Federals in flank the next day at Glendale on the Quaker Road; and for Longstreet (with A. P. Hill under his command), with 18,000 men, to march down to Darbytown Road, a couple of miles south of Charles City Road, then turn eastward and also attack McClellan around Glendale on June 30. Meanwhile, Holmes was to press eastward along River Road, and Jackson was to rebuild bridges over the Chickahominy, then move eastward and intercept McClellan if, by chance, he turned toward Fort Monroe.

16. Magruder was slow and allowed himself to be stopped by a single Union brigade at Savage Station, three miles east of Seven Pines, permitting Union troops to get away under cover of night. Huger showed great hesitation and advanced only six miles down Charles City Road before nightfall. Holmes likewise proceeded only a short distance. Longstreet did better, reaching within three miles of Glendale by dark. It took Jackson's men nearly all day to rebuild two bridges on the Chickahominy, and they did not cross until after midnight.

17. Jackson's biographer James I. Robertson Jr. gives a detailed account of Jackson's halt at White Oak Bridge. Jackson, along with his cavalry chief, T. T. Munford, crossed the stream with horses but returned when they found cavalry

could not operate alone because of sharpshooters and cannons. A. R. "Rans" Wright found a crossing a short distance upstream at Fisher's Ford. He established a bridgehead there but failed to inform Jackson. Wade Hampton found a second possible crossing a little east of the broken bridge, but it was suitable only for infantry. If infantry alone were able to cross, they would have been helpless, because artillery and supply wagons could not go with them. Robertson says that the situation reflected an almost complete breakdown in communications between the generals on June 30. See Robertson, 493–97.

18. It was not for Jackson to initiate a roundabout movement to the Charles City Road. Lee's orders to Jackson were to move along the main road to White Oak Bridge, to endeavor to force his way toward Glendale, and to guard the army's left flank, if by chance McClellan had set up an attack across the fords or bridges of the lower Chickahominy. He received no further orders from Lee. On July 13, 1862, Stapleton Crutchfield, Jackson's artillery chief; Sandie Pendleton, his aide; and Hunter McGuire, his medical officer, were discussing whether it would have been better for Jackson to have gone to Longstreet's aid. Jackson came in while the discussion was going on, and said curtly: "If General Lee had wanted me he could have sent for me." But no such order was sent by Lee. A march of about six miles over little-used but accessible country lanes and a ford across the swamp would have put troops on the Charles City Road at White's Tavern, four and a half miles west of Glendale. (Source: personal letter of Hunter McGuire to G. F. R. Henderson. See Henderson, vol. 2, 57–58.)

19. Much about the Seven Days, such as Jackson's behavior at White Oak Bridge, is inexplicable. Lee's failure to recognize the immense bonanza presented at Malvern Hill on June 30 is another mystery. The situation at Glendale was already proving to be a stalemate, since Jackson and Huger were evidently not going to push the Union forces from behind. Yet Lee had a golden opportunity to occupy Malvern Hill and produce precisely the same effect—blocking the enemy's retreat—that he was seeking at Glendale. Since Malvern Hill was being held at the time mainly by artillery unprotected by infantry, it could have been quickly occupied with infantry armed with the Minié-ball rifle. It is clear that Lee saw this opportunity, for he directed Holmes to seize the hill. Yet he knew Holmes was an untried field commander, and entrusting the task to him was an enormous gamble. In the event, Lee failed at both Glendale and Malvern Hill to block the enemy.

20. If Magruder and Holmes, neither more than thirteen miles away, had been ordered to join Longstreet and A. P. Hill at Glendale, and a part of Huger's division brought over, Lee could have assembled a very powerful force for an attack, and the enemy might have been defeated by sheer power, as Porter was overwhelmed at Gaines Mill.

21. Porter Alexander, *Military Memoirs,* 167–71; Henderson, vol. 2, 70–71; Walter Taylor, 41–44.

22. Johnson and Buel, vol. 2, 391.

23. *War of the Rebellion,* vol. 11, part 2, 619.

24. Johnson and Buel, vol. 2, 432; Porter Alexander, *Fighting for the Confederacy,* 114; Freeman, *R. E. Lee,* vol. 2, 220.

25. McClellan rode around Malvern Hill early on Tuesday morning, July 1, but abandoned the field by 9:15 A.M., climbed back on the *Galena,* and steamed to Harrison's Landing. Porter took over as commander by default. McClellan returned at about 3:30 P.M., but he went only to the extreme right, well away from the scene of the fighting, which he played no role in directing.

26. *War of the Rebellion,* vol. 11, part 3, 282.

Chapter 5: The Sweep Behind Pope

1. Dabney, 486–87; Henderson, vol. 2, 76–78; *Southern Historical Society Papers,* vol. 40, 180–82. Porter Alexander shared Jackson's view. Lee, he says, could not wait idly at Richmond for the enemy slowly to make up his mind. See Porter Alexander, *Military Memoirs,* 179.

2. *Southern Historical Society Papers,* vol. 25, 110, 119; Heros von Borcke, *Memoirs of the Confederate War for Independence* (New York: Peter Smith, 1938), vol. 2, 117.

3. Luring Pope to attack and leaving an open flank around which Longstreet could swing after Pope had been defeated is precisely what Jackson arranged for in the battle of Second Manassas in late August 1862. It is also what he tried, and failed, to induce Lee to do to destroy Ambrose E. Burnside's army in December 1862.

4. This actually happened on two occasions: John Pope left the eastern flank of his army exposed on the Rapidan River in August 1862, and Joseph Hooker left the western flank of his army exposed at Chancellorsville in May 1863.

5. *Southern Historical Society Papers,* vol. 25, 119; Bigelow, 340n. Jackson's close associate, Robert Lewis Dabney, expressed the same idea in somewhat different words. See Dabney, 699–700.

6. Lincoln appointed Pope on June 27, 1862, to take command of the forces of Frémont, Banks, and McDowell, totaling 50,000 men, plus 12,000 garrisoning Washington, as well as Brigadier General Jacob D. Cox, with 11,000 men, coming from West Virginia. See Johnson and Buel, vol. 2, 281 and 451; Allan, *Army of Northern Virginia,* 152. On July 11 Lincoln named Halleck as commander in chief of all Federal armies. Halleck arrived in Washington on July 23. See *War of the Rebellion,* vol. 11, part 3, 314; Allan, *Army of Northern Virginia,* 154; Porter Alexander, *Military Memoirs,* 177. Lincoln wanted Pope to draw off Confederate troops and relieve pressure on McClellan. He named Halleck to bring about cooperation between McClellan and Pope. Pope's mission, as outlined on June 26, 1862, was to cover Washington, control the Shenandoah, "and at the same time so operate

upon the enemy's lines of communication in the direction of Gordonsville and Charlottesville as to draw off, if possible, a considerable force of the enemy from Richmond and thus relieve the operations against that city of the Army of the Potomac." See Johnson and Buel, vol. 2, 449–50.

7. John Pope brought a new level of horror to the war by instructing his army to live off the country and reimburse only citizens loyal to the United States, and to hold local citizens hostage for actions of guerrillas, without distinguishing between partisans and ordinary Confederate cavalry. He went so far as to treat a mother who wrote her soldier son as a traitor, subject to being shot as a spy. Pope's orders so incensed Lee that he wrote twice that Pope must be suppressed. Pope was the only adversary whom Lee regarded with undisguised contempt. With the approval of President Davis, Lee notified Halleck that the Confederacy would be compelled to retaliate if Pope's orders were enforced. Halleck modified Pope's orders materially but did not renounce them entirely. See Cooke, 252–54; Henderson, vol. 2, 81–82.

8. Lee, 56–58; Freeman, *R. E. Lee,* vol. 2, 259, 328–29.

9. Porter Alexander, *Military Memoirs,* 179.

10. Pope was the most ridiculed and vilified of the Union generals who faced the Army of Northern Virginia. His reputation was scarcely better in the Federal army. Pope made the mistake of issuing a bombastic order to his troops on July 14, 1862, from his "headquarters in the saddle." In it he said: "I come to you from the West where we have always seen the backs of our enemies; from an army whose business it has been to seek the adversary and beat him when he was found; whose policy has been attack and not defense. . . . I presume I have been called here to pursue the same system and to lead you against the enemy. . . . Meantime, I desire to dismiss from your minds certain phrases, which I am sorry to find so much in vogue amongst you. I heard constantly of 'taking strong positions and holding them'; of 'lines of retreat' and of 'bases of supplies.' Let us discard such ideas. . . . Let us study the possible lines of retreat of our opponents and leave our own to take care of themselves. . . . Success and glory are in the advance. Disaster and shame lurk in the rear." General Lee reviled the general and said of his own nephew Colonel Lewis Marshall, who served on the staff of Union General Nathaniel P. Banks: "I could forgive [his] fighting against us but not his joining Pope." See *War of the Rebellion,* vol. 12, part 3, 474; Porter Alexander, *Military Memoirs,* 176 -77; Allan, *Army of Northern Virginia,* 156–58; Freeman, *R. E. Lee,* vol. 2, 263–64.

11. A. P. Hill's division had suffered 4,210 casualties in the Seven Days and had been reduced to 10,623 men. However, the Louisiana brigade of Brigadier General Leroy A. Stafford was attached, raising his force to about 12,000 men. See Allan, *Army of Northern Virginia,* 165; Porter Alexander, *Military Memoirs,* 174. Lee advised Jackson subtly that he should confide more in his subordinate commanders,

especially the capable but impetuous and trigger-tempered A. P. Hill. But Jackson's habit of secrecy was too ingrained, and he ignored the good advice. See *War of the Rebellion,* vol. 12, part 3, 919; Freeman, *Lee's Lieutenants,* vol. 2, 3.

12. Henderson, vol. 2, 83.

13. Johnson and Buel, vol. 2, 281.

14. In Jackson's plan to seize Culpeper one can see a silhouette of the kind of strategy he might have followed if Lee and Davis had released enough men to undertake an offensive campaign against Pope. Jackson likely would have advanced to a central position between the still-separated parts of Pope's army and tried to destroy each part one at a time.

15. Freeman, *Lee's Lieutenants,* vol. 2, 20; Allan, *Army of Northern Virginia,* 167, 169. Banks's corps totaled 14,500 infantry and artillerymen and 4,100 cavalry. But he left 2,500 infantry and gunners at Winchester and 1,000 at Front Royal, while sickness and heat prevented Banks from bringing more than 9,000 men to Cedar Mountain.

16. The day after the battle of Cedar Mountain, Sigel's corps and 5,000 more cavalry joined Ricketts's division and Banks's shaky troops, giving Pope 32,000 men in place, while Rufus King's division was marching from Fredericksburg, and other detachments were coming up. Shortly after the battle Jesse L. Reno's corps, 8,000 men, from Burnside's force also arrived. See *War of the Rebellion,* vol. 12, part 2, 185; Henderson, vol. 2, 106–7; Porter Alexander, *Military Memoirs,* 182, 185.

17. Pope massed his main body on the Gordonsville-Culpeper road, well to the west of the eastern fords over the Rapidan. Aware that Jackson was based on Gordonsville, Pope was convinced that if Jackson advanced, he would come directly up the road to Culpeper. He never even considered that the Confederates might go around him. See Henderson, vol. 2, 113.

18. *War of the Rebellion,* vol. 12, part 2, 725–29; part 3, 950–41; Henderson, vol. 2, 111–12; Porter Alexander, *Military Memoirs,* 186–89; Freeman, *R. E. Lee,* vol. 2, 259, 282.

19. Freeman, *R. E. Lee,* vol. 2, 284.

20. This illustrated the old saw that for the want of a nail a nation was lost. When no cavalry appeared at Verdiersville as expected on the night of August 17, General Longstreet ordered two infantry regiments to be put on picket duty on the road to Raccoon Ford. The command went to the brigade commanded by Georgia politician-turned-brigadier Robert Toombs. He, however, was visiting a neighboring brigadier. The senior colonel sent out the two regiments and they were duly posted to guard the ford. Toombs came along soon afterward, found the regiments were part of his brigade, and angrily ordered them back to camp, claiming that no orders should be obeyed from superior officers that did not come through him. Thus the Federal cavalry patrol got within Confederate lines undetected, the strike by Confederate cavalry did not come off, the offensive was fatally delayed, and the

chance to destroy Pope's army was forfeited. When the facts came out, one of Longstreet's adjutants, dressed formally in sword and sash, arrived at Toombs's brigade and placed the brigadier under arrest. Toombs at last apologized, was restored to duty, and rejoined his brigade in time to take part in the battle of Second Manassas on August 30. See Porter Alexander, *Military Memoirs,* 188.

21. Porter Alexander, *Military Memoirs,* 187–88; Henderson, vol. 2, 114; Freeman, *Lee's Lieutenants,* vol. 2, 57–61; Freeman, *R. E. Lee,* vol. 2, 284–86.

22. Dabney, 511–12; Henderson, vol. 2, 114–15; Allan, *Army of Northern Virginia,* 182.

23. George H. Gordon, *The Army of Virginia* (Boston: Houghton, Osgood and Company, 1880), 9.

24. Arriving at Jackson's headquarters at Jeffersonton on August 24, Jeb Stuart displayed a beautiful blue uniform coat, inside of which was a tag with the name of its owner: "John Pope, Major General." He had seized it in the raid of Pope's headquarters at Catlett Station. Stuart announced he had a proposition to make. Taking a piece of paper, he wrote "Major Genl. John Pope, Commanding, etc. General: You have my hat and plume. I have your best coat. I have the honor to propose a cartel for a fair exchange of the prisoners. Very Respectfully, J. E. B. Stuart, Maj. Genl., C.S.A." Stuart had been enduring endless ribbing—from privates to generals—for having lost his hat and plume to Union cavalrymen in his hasty flight from Verdiersville on August 17. The proposed "cartel" amused Jackson greatly. The communication was sent through the lines. Nothing ever came of it. Perhaps it did not amuse the other party as much as it did Stuart and the other Confederates. See Douglas, 133–34.

25. Fitz John Porter's and Samuel P. Heintzelman's corps from McClellan's army, plus John F. Reynolds's Pennsylvania division—in all, 25,000 men—were within two days' juncture with Pope, while Jacob D. Cox, with 7,000 of his force from West Virginia, and Samuel D. Sturgis, with 10,000 from Washington, along with the remainder of McClellan's army, were not more than five to eight days away. Pope soon would possess more than 120,000 troops.

26. Historians later criticized Lee for this move because he divided his forces in the face of the enemy; it left him with 32,000 men until D. H. Hill's and McLaws's divisions could arrive. Lee responded that such criticism was obvious but that the disparity of force between the contending armies rendered the risks unavoidable. See Allan, *Army of Northern Virginia,* 200n.

27. Although Hunter McGuire, Jackson's surgeon, believed that Jackson had originated the idea for the movement on Pope's rear (Henderson, vol. 2, 123–24), it is evident that the proposal came from Lee. See Freeman, *R. E. Lee,* vol. 2, 300–1; Freeman, *Lee's Lieutenants,* vol. 2, 82–83; *War of the Rebellion,* vol. 12, part 2, 553–54 and 642–43; and Porter Alexander, *Military Memoirs,* 191–92. The dif-

ference between Lee's and Jackson's concepts of the mission is shown in the small but significant disparities between Lee's and Jackson's accounts of the operation (see *War of the Rebellion* citations above). Lee's report reads: "In pursuance of the plans of operations determined upon, Jackson was directed on the 25th [of August 1862] to cross above Waterloo and move around the enemy's right, so as to strike the Orange and Alexandria Railroad in his rear." Jackson's report reads: "Pursuing the instructions of the commanding general, I left Jeffersonton on the morning of the 25th to throw my command between Washington City and the army of General Pope and to break up his railroad communication with the Federal capital." Lee saw the movement as a way to cut the railway. Jackson saw it as a deep strategic strike aimed at Pope's connection with the capital and, by inference, at the capital itself, an immensely greater and more pivotal move. Jackson was seeking a decision in the war; Lee was seeking to evict Pope from the Rappahannock.

28. Near Salem, Jackson climbed on a large rock and, taking off his hat, watched the soldiers march by. A message swept down the ranks: "No cheering, boys; the General requests it." The soldiers passed by silently but showed their affection by waves and swinging their caps and hats in the air. Jackson turned to his staff, his face beaming. "Who could not conquer with such troops as these?" Dabney, 517; Henderson, vol. 2, 126–27.

29. Allan, *Army of Northern Virginia,* 208–9; *War of the Rebellion,* vol. 12, part 2, 333. Pope (Johnson and Buel, vol. 2, 460–61), writing some twenty years after the war, attempted to hide the fact that he made no response to Jackson's march on his rear. "Stonewall Jackson's movement on Manassas Junction was plainly seen and promptly reported, and I notified General Halleck of it," he wrote. He tried to pass off his responsibility to unnamed others by saying he had directed some of his forces en route to him to meet this threat, but "all of the reinforcements and movements of the troops promised me had altogether failed." However, the editors of *Century* magazine, which published the series, pointed out in a footnote that Pope's orders for August 25 disposed his troops on the line of the Rappahannock, as for an advance toward the Rapidan, which indicates he was totally unaware of what Jackson was doing.

30. Allan, *Army of Northern Virginia,* 209.

31. Ibid., 211; Johnson and Buel, vol. 2, 461, 463, 517; *War of the Rebellion,* vol. 12, part 3, 672, 675, 684; Henderson, vol. 2, 138. Pope asked Halleck that William B. Franklin's 10,000-man corps of McClellan's army, being directed to Alexandria rather than Aquia Creek, be sent to Gainesville. He also asked that another division, either Jacob D. Cox's from West Virginia or Samuel D. Sturgis's from Washington, be moved to Manassas Junction. Halleck sent none of these forces forward, however.

32. Johnson and Buel, vol. 2, 505. Cutting the telegraph line severed Pope's

direct connections with Washington from 8:20 P.M. August 26 until August 30. During this period all communications had to go via Falmouth, just north of Fredericksburg. See Johnson and Buel, vol. 2, 461n.; Allan, *Army of Northern Virginia*, 214.

33. English strategist Major General J. F. C. Fuller (Fuller, *Grant and Lee*, 165) criticizes Lee for not striking at once for Berryville in the Shenandoah Valley, and calling for Jackson to join him by way of Snicker's Gap (on present-day Virginia Route 7), then advancing with the concentrated army on Harpers Ferry. From there Lee could have threatened Washington. This would have compelled Halleck and Pope to withdraw all forces from Virginia to defend the capital, and would have avoided a battle. Fuller finds it astonishing that Lee abandoned the strategic offensive by moving to join Jackson, and thereby assumed the tactical offensive. Fuller's criticism of Lee is valid, since Lee's stated purpose (to President Davis) was to avoid a battle and merely to maneuver Pope out of Virginia. Apparently this solution did not occur to Lee, however. Stonewall Jackson, on the other hand, was not in the least interested in evicting Pope from Virginia. He wanted to destroy Pope's army. Thus his motivation in moving to Manassas Junction was to *provoke* Pope into trying to destroy his force, thereby challenging Jackson to battle. The whole purpose of Jackson's subsequent moves, therefore, was to deceive Pope into coming after him.

34. James Longstreet (Johnson and Buel, vol. 2, 517) reported that until late on August 28 "we had received reports from General Jackson, at regular intervals, assuring us of his successful operations and of confidence in his ability to baffle all efforts of the enemy till we should reach him."

35. *War of the Rebellion*, vol. 12, part 2, 71–72; Allan, *Army of Northern Virginia*, 221–23.

Chapter 6: Second Manassas

1. In *War of the Rebellion* (vol. 12, part 2, 644–45), Jackson writes: "As he [the enemy] did not appear to advance in force and there was reason to believe the main body was leaving the road and inclining toward Manassas Junction, my command was advanced through the woods, leaving Groveton on the left, until it reached a commanding position near Browner's house," three-quarters of a mile west-northwest of Groveton crossroads.

2. Johnson and Buel, vol. 2, 470–71, 517–18; *War of the Rebellion*, vol. 12, part 2, 393; Allan, *Army of Northern Virginia*, 241–42.

3. Pope justifiably complains in Johnson and Buel (vol. 2, 470–71) regarding the absence of McDowell and the failure of leadership and resolve of all his division and corps commanders in this sector at the time. Pope's orders during the night of August 28 and early morning of August 29 indicate that he sincerely be-

lieved that his left, or western, flank was preventing the junction of Lee and Jackson, and thus he could concentrate the remainder of his army against Jackson.

4. Henderson, vol. 2, 152.

5. Ropes, 85–86.

6. Johnson and Buel, vol. 2, 473–74; *War of the Rebellion,* vol. 12, part 2, 518, 520; Allan, *Army of Northern Virginia,* 257–58; Porter Alexander, *Military Memoirs,* 204.

7. Sigel's corps had about 13,000 men, Reynolds's division 8,000. The attack therefore was no stronger than the defense, and the Federal artillery was restricted from giving much support because of the woods. Even these forces attacked piecemeal or did not fully engage. When Schurz attacked the Confederate left, the brigade of Robert H. Milroy and the division of Robert C. Schenck advanced over the open meadows to the left of the woods, while Reynolds moved forward on their left. Adolph von Steinwehr's division remained in reserve. When Schurz's troops were shattered, none of the other Federal forces engaged. This was the pattern of attacks for the whole day. See Henderson, vol. 2, 155–58.

8. After the battle Pope court-martialed Porter. When the court finished its deliberations months later, it cashiered Porter. It was not until 1882, after a board had largely vindicated Porter, that President Chester Arthur remitted Porter's sentence.

9. Johnson and Buel, vol. 2, 481.

10. There was an enigmatic letter written by Longstreet to Fitz John Porter in regard to Porter's efforts to get his court-martial overturned (he was court-martialed primarily for failing to attack Longstreet on August 29). In this letter Longstreet wrote, concerning the hours in which his command came up to the west of Jackson: "We all were particularly anxious to bring on the battle after twelve M. [*meridies,* or noon], General Lee more so than the rest." However, according to William Allan, at midday on August 29 "General Lee was ready to *receive* attack at all points of his lines, and it seemed that a general battle was about to be joined." Allan says that "Lee preferred to receive the attack rather than give it." He refers to a letter from Lee to Porter on February 18, 1870: "The result of an attack [by Porter on Longstreet after noon on August 29] would have been a repulse, and if a repulse, especially at an early hour or before 5 P.M., the effect would have been an attack on General Pope's left and rear by Longstreet and Stuart, which, if successful, would have resulted in the relief of Jackson, and have probably rendered unnecessary the battle the next day." See Allan, *Army of Northern Virginia,* 260–61. The conclusion one can draw is that Lee did *not* want to attack on August 29, and that had Jackson not provoked Pope into an engagement, Lee would have made no effort to challenge him to a battle.

11. Maurice, 142–43.

12. Irvin McDowell and Samuel Peter Heintzelman, after a personal reconnaissance of their own, apparently decided Jackson was departing not only because his lines were quiet but also because there were no Rebel cavalry north of Bull Run, as had been the case the day before. But the cavalry were there then because Fitz Lee returned on the 29th from raids on Burke station and Fairfax Court House by way of Centreville and Sudley.

13. *War of the Rebellion,* vol. 12, part 3, 741; Allan, *Army of Northern Virginia,* 267; Freeman, *R. E. Lee,* vol. 2, 329. After the battle in his reports, Pope tried to hide the fact that he had actively sought to attack, claiming that he had been trying to "hold my position" and delay any "farther advance" of the Rebels "toward the capital," and that his plans had not involved a general assault on what he termed "superior forces" of the enemy. See Johnson and Buel, vol. 2, 486; *War of the Rebellion,* vol. 12, part 2, 41. But as William Allan (*Army of Northern Virginia,* 266) writes, "To lead an inferior and exhausted army in hopeless struggle against a well-posted adversary in order to gain time was not the real part played by General Pope on that memorable day, however fully he may afterwards have persuaded himself that such was the case."

14. *War of the Rebellion,* vol. 12, part 2, 361, 384, 413; Allan, *Army of Northern Virginia,* 270, 276.

15. Porter Alexander, *Fighting for the Confederacy,* 251.

16. *War of the Rebellion,* vol. 12, part 2, 394; Allan, *Army of Northern Virginia,* 286; Porter Alexander, *Military Memoirs,* 213–14; Henderson, vol. 2, 173.

17. *War of the Rebellion,* vol. 12, part 2, 666–67; Allan, *Army of Northern Virginia,* 282; Henderson, vol. 2, 175.

18. When Kearny's widow applied for Kearny's mount and horse furnishings captured when he fell, Lee had them appraised, paid for them himself, and sent them to Mrs. Kearny. See Freeman, *R. E. Lee,* vol. 2, 429–33; *War of the Rebellion,* vol. 19, part 2, 645, 654–55.

Chapter 7: The Lost Order

1. This is proved by his documented plea to Lee to meet the Union army east of Frederick, Maryland (Dabney, 548–49); by the strong defensive position with an open flank he set up at Groveton in the battle of Second Manassas, which was the model of the kind of battle he wanted to fight; by his urging Lee to swing around the unguarded eastern flank of John Pope's army on the Rapidan River; and by his subsequent unsuccessful effort to convince Lee not to fight a wholly defensive battle at Fredericksburg in December 1862, but instead to withdraw twenty-five miles south to the North Anna River, where the Confederates could defeat any attack and also have space to swing around the flanks of the demoralized Union army. All of these were consistent with the defend-then-attack strategy that

Jackson developed after the Seven Days. They were a direct challenge to Lee's demonstrated fixation on frontal assaults.

2. On September 8, 1862, Lee told division commander John G. Walker that breaking the bridge at Harrisburg would mean that, since the Confederates would cut the B&O Railroad with their movement north, "there will remain to the enemy but one route of communication with the west, and that very circuitous, by way of the [railroads near the Great] Lakes. After that, I can turn my attention to Philadelphia, Baltimore, or Washington, as may seem best for our interests." See Johnson and Buel, vol. 2, 604–6; *War of the Rebellion,* vol. 19, part 2, 592; Freeman, *Lee's Lieutenants,* vol. 2, 160 n. 30; *R. E. Lee,* vol. 2, 362.

3. Upon reaching Frederick, Maryland, Lee assembled all his leading generals in a council of war to devise a plan of operations. It was here that Jackson expressed his opposition to Lee's plan and proposed his alternative. See R. L. Dabney's account (Dabney, 548–49), as well as G. F. R. Henderson's (vol. 2, 212).

4. Johnson and Buel, vol. 2, 554.

5. *War of the Rebellion,* vol. 19, part 1, 65.

6. Freeman, *Lee's Lieutenants,* vol. 2, 721.

7. Dabney, 548–49. See also Henderson, vol. 2, 212. Dabney was a noted Presbyterian theologian at Hampden-Sydney College, near Farmville, Virginia, and a longtime friend of Jackson. Jackson recruited him as his chief of staff (adjutant general) in March 1862 with the rank of major. Dabney served through the Valley campaign and the Seven Days, but, because of ill health, was returned to civilian life in July 1862. He remained a close confidant of Jackson, and published his book on Jackson's campaigns in 1866, only a year after the end of the war.

8. Henderson, vol. 2, 212. Henderson reported that Longstreet objected to dividing the army to capture Harpers Ferry. Longstreet told Lee that if the army was kept together and provided with supplies, "then we could do anything we pleased." Longstreet added that "General Lee made no reply to this, and I supposed the Harpers Ferry scheme was abandoned." Henderson wrote: "Jackson would have preferred to fight McClellan first [in front of Frederick], and consider the question of communications afterwards." In this case "communications" means reduction of Harpers Ferry to ensure the supply line through the Shenandoah Valley. In other words, Jackson wanted to confront McClellan somewhere around Frederick and then worry about a supply line, the implication being that McClellan would lose and that supplies then would take care of themselves.

9. *War of the Rebellion,* vol. 19, part 2, 600; Freeman, *R. E. Lee,* vol. 2, 358. President Davis made no response to Lee's letter. The president's silence was all the authorization Lee sought. Davis probably had a sense of the true meaning of Lee's move. As Steven E. Woodworth writes, "This move smelled of a showdown fight." There was no possibility that the Lincoln administration would entertain any peace proposal without an appeal to battle. See Wentworth, 185–88.

10. The Confederate Quartermaster Department tried to pass off blame for shortages onto unit commanders. The quartermaster general, Abraham C. Myers, complained that field supply officers failed to notify him of shortages. Instead, Lee wrote directly to the secretary of war, George W. Randolph, or President Davis after conditions became critical. But it was evident that the Quartermaster Department failed to deliver goods to the troops that were actually on hand. There was far less excuse for food shortages, other than difficulties of transportation. The Commissary Department had accumulated ample food stocks. But they were often not delivered. See Richard D. Goff, *Confederate Supply* (Durham, N.C.: Duke University Press, 1969), 67–68, 76–78.

11. *War of the Rebellion,* vol. 19, part 2, 590; Allan, *Army of Northern Virginia,* 323; Woodworth, 185.

12. Johnson and Buel, vol. 2, 687–88.

13. *War of the Rebellion,* vol. 19, part 2, 590; Allan, *Army of Northern Virginia,* 323; Woodworth, 185.

14. Leighton Parks, *Century* magazine, vol. 70, number 2, 255ff.; Freeman, *R. E. Lee,* vol. 2, 355.

15. *War of the Rebellion,* vol. 19, part 1, 145; part 2, 603–4; Allan, *Army of Northern Virginia,* 332–33; Freeman, *R. E. Lee,* vol. 2, 363.

16. Kyd Douglas describes the roundabout march by Jackson's corps. Just before departing, Jackson asked for a map of the Pennsylvania frontier and inquired as to roads and localities north of Frederick. These questions went out into the community, as Jackson intended, causing spies to conclude that the Confederate army was heading into Pennsylvania. Douglas writes (Douglas, 152): "The cavalry, which preceded the column, had instructions to let no civilian go to the front, and we entered each village we passed before the inhabitants knew of our coming. In Middletown two very pretty girls, with ribbons of red, white, and blue in their hair, and small Union flags in their hands, came out of a house as we passed, ran to the curbstone, and laughingly waved their colors defiantly in the face of the general. He bowed and lifted his cap and with a quiet smile said to the staff, 'We evidently have no friends in this town.' "

17. Porter Alexander, *Fighting for the Confederacy,* 141. D. H. Hill was extremely sensitive about the lost order and maintained that it helped Lee, since McClellan thought Longstreet's force was at Boonsboro, not Hagerstown, therefore making him more cautious. See *Southern Historical Society Papers,* vol. 13, 420–21. Lee, on the other hand, was convinced that the order permitted McClellan to move quickly against him, when without it he would not have done so.

18. *War of the Rebellion,* vol. 19, part 1, 951. During the night of September 14 more than a thousand Union cavalrymen escaped by crossing the pontoon bridge into Maryland and moving northwest, capturing forty-five wagons of Longstreet's

ordnance train as the cavalry were fleeing into Pennsylvania. The Union cavalry took a road around the western base of Maryland Heights that McLaws had failed to block. Colonel Miles also could have evacuated a large part of the Federal infantry by this same unguarded road if he had shown any initiative. McLaws's failure to post troops on this road shows how the mistake of a subordinate can jeopardize even the best plans of an army commander. The error probably contributed to Lee's refusal afterward to give McLaws wider responsibilities. See Sears, 151; Porter Alexander, *Fighting for the Confederacy,* 144.

19. Porter Alexander, *Fighting for the Confederacy,* 146. Alexander on p. 147 gives a sketch map of the difficult approach to the ford and the difficult passage from the ford.

20. Freeman, *R. E. Lee,* vol. 2, 411; Freeman, *Lee's Lieutenants,* vol. 2, 225. Garnet Wolseley, later commander in chief of the British army, reported that Lee told him in October 1862 that he had no more than 35,000 men at Antietam. See Wolseley, 21. In an interview with William Allan at Lexington, Virginia, on February 15, 1868 (Freeman, *Lee's Lieutenants,* vol. 2, 721), Lee said he had about 35,000 men in the battle. See also Cooke, 340–41; Porter Alexander, *Military Memoirs,* 244–45. For an analysis of the figures in the battle, see Allan, *Army of Northern Virginia,* 379, 397–98; *War of the Rebellion,* vol. 19, part 1, 67; Henderson, vol. 2, 242–43. Johnson and Buel (vol. 2, 603) give total Federal strength as 87,164 and Confederate as "less than 40,000." The Army of the Potomac actually counted 94,000 troops, but this figure included cooks, teamsters, and other detailed men. In addition, a 6,600-man division of Porter's corps was at Frederick, and marching hard to the battlefield.

21. Allan, *Army of Northern Virginia,* 440–41; Freeman, *Lee's Lieutenants,* vol. 2, 715–23. See also *War of the Rebellion,* vol. 19, part 2, 592; Johnson and Buel, vol. 2, 604–5.

22. James Longstreet (Johnson and Buel, vol. 2, 666–67) says that Lee should have retired from Sharpsburg the day he learned that Jackson had captured Harpers Ferry. He adds: "The moral effect of our move into Maryland had been lost by our discomfiture at South Mountain and it was then evident we could not hope to concentrate in time to do more than make a respectable retreat, whereas by retiring before the battle we could have claimed a very successful campaign." Porter Alexander (*Fighting for the Confederacy,* 145–46) calls the decision to stand at Antietam "the greatest military blunder that General Lee ever made." A drawn battle was the best possible outcome one could hope for, he writes. See also Porter Alexander, *Military Memoirs,* 242–49. Major General Sir Frederick Maurice (Maurice, 152) says: "Of all Lee's actions in the war this seems to me to be the most open to criticism. He was only justified in giving battle if retreat was impossible without fighting or if he had a good prospect, not merely of repulsing attacks, but

of beating his enemy soundly. . . . The ground he chose for battle, while admirably suited for defense, left him no opportunity for such a counterstroke as Longstreet had delivered at the second battle of Manassas. He could at best hope to beat off the Federals. But at the end of a battle he would be no better off than he was on the morning of the 15th. The Antietam, the most desperately fought struggle of the war, must be numbered among the unnecessary battles." Major General J. F. C. Fuller (Fuller, *Grant and Lee,* 169) calls Antietam "a totally unnecessary battle," caused, he says, because Lee's "personal pride could not stomach the idea that such an enemy [as McClellan] could drive him out of Maryland."

23. Given the fact that the South was ruled by its landed aristocracy, there was no way the Confederate army's command structure could have been altered, no matter how much acclaim Jackson might receive. An example of the outlook of the Southern ruling class was shown when Jackson and Lee met with President Davis at the presidential mansion on Clay Street in Richmond on July 13, 1862, just prior to Jackson's departure for Gordonsville in the campaign against John Pope. After the conference ended, Lee and Jackson tarried on the front steps of the mansion. Captain Charles Blackford, of the 2nd Virginia Cavalry, just attached to Jackson's staff, saw the two generals side by side for the first time. He was struck by the contrast in their appearance. Blackford wrote: "Lee was elegantly dressed in full uniform, sword and sash, spotless boots, beautiful spurs, and by far the most magnificent man I ever saw. The highest type of the Cavalier class to which by blood and rearing he belongs. Jackson, on the other hand, was a typical Round-head. He was poorly dressed, that is, he looked so though his clothes were made of good material. His cap was very indifferent and pulled down over one eye, much stained by weather and without insignia. His coat was closely buttoned up to his chin and had upon the collar the stars and wreath of a general. His shoulders were stooped. . . . He had a plain swordbelt without sash and a sword in no respect different from that of other infantry officers that I could see. His face, in repose, is not handsome or agreeable and he would be passed by anyone without a second look." See Robertson, 511–12; Susan Leigh Blackford, comp., *Letters from Lee's Army* (New York: 1947), 86. Captain Blackford disdained Jackson's headquarters as "nothing but a roll of blankets strapped up and two campstools and a table." He was present at the battle of Cedar Mountain when Jackson rode in to rally the Confederate soldiers thrown into confusion by the advance of Banks's right wing. Blackford wrote: "Jackson usually is an indifferent and slouchy looking man but then, with the 'Light of Battle' shedding its radiance over him his whole person had changed. . . . Even the old sorrel horse seemed endowed with the style and form of an Arabian." At one of Jackson's supreme moments, social arbiter Blackford felt Stonewall should be seen as being mounted on a more stylish horse than he actually was. See Robertson, 517, 532.

24. Douglas, 166.

Chapter 8: Antietam

1. Lee's letter to Mrs. Jackson is in the one-volume reprint of Henderson's book, *Stonewall Jackson and the American Civil War* (New York: Longmans, Green and Co., 1936, 1937, 1943, 1949), 694–95. Dabney's version is in Dabney, 570. Robertson, 608, says the Dabney letter is in the Robert Lewis Dabney Papers at Union Theological Seminary, Richmond, and Lee's letter to Anna Jackson is in the George and Catherine Davis Collection at Tulane University, New Orleans.

2. Details of the battle of Antietam are drawn from the following sources: Allan, *Army of Northern Virginia,* 372–437; Sears, 168–297; Henderson, vol. 2, 239–88; Robertson, 584–623; Freeman, *R. E. Lee,* vol. 2, 382–404; Freeman, *Lee's Lieutenants,* vol. 2, 203–25; Johnson and Buel, vol. 2, 596–603, 627–02, 667–72, 675–85; Douglas, 168–70, Cooke, 327–42; Palfrey, 57–137; Chambers, vol. 2, 218–34; John M. Priest, *Antietam: The Soldiers' Battle* (Shippensburg, Pa.: White Mane, 1989).

3. Once the enemy had stripped forces from a point to defend against the flank movement, Napoleon carried out a decisive third step to ensure victory: he sent a powerful force directly into the weakened enemy point to create a breakthrough. This final coup de grâce would not have been necessary for McClellan, since Lee's army was so inferior in size that it could not have withstood a strong turning movement to the south.

4. Theodore A. Dodge, *Alexander* (New York: Houghton Mifflin, 1890; London: Greenhill, 1991), 657.

5. An alternative tactical plan that could have worked at Sharpsburg was closer to what McClellan himself proposed but did not carry out: hard attacks on both the north and the south, pulling Lee's reserves to these points, then a massive stroke into the center of the Confederate position along Boonsboro Pike to split the Southern army in half. This would have been harder to pull off, since it would have consisted entirely of frontal attacks. But it would probably have resulted in Lee's defeat, although that part of the army to the south might have been able to get across the Potomac.

6. Porter Alexander, *Fighting for the Confederacy,* 146.

7. Although the Union's initial attack hit the Rebels' northern flank and was described by McClellan as a flank attack, tactically it was a frontal assault against Lee's northern face. Jackson aligned his corps to defend against this attack. This had the effect of bending the line in Jackson's sector to front largely north, not east, as did the remainder of the line. Lee had so few troops that on the extreme northern flank he was able to post only Jeb Stuart's cavalry and some guns, protected by a single infantry brigade (Jubal Early's). Although this force was a nuisance to the Federals, it was incapable of flanking the Union attack that Jackson saw was coming directly from the north, down the Hagerstown Pike.

8. On the morning of September 18, 1862, Lee and Jackson were still discussing whether they could outflank McClellan's army. Lee told artillery colonel Stephen D. Lee to report to General Jackson. Jackson took Colonel Lee to a hill on the north of the Confederate line. Stephen Lee wrote: "General Jackson said: 'Colonel, I wish you to take your glasses and carefully examine the Federal line of battle.' I did so, and saw a remarkably strong line of battle, with more troops than I knew General Lee had. After locating the different batteries . . . I said to him, 'General, that is a very strong position, and there is a large force there.' He said: 'Yes. I wish you to take fifty pieces of artillery and crush that force, which is the Federal right. Can you do it?' . . . I at once saw such an attempt must fail. More than fifty [Federal] guns were unlimbered and ready for action, strongly supported by dense lines of infantry and strong skirmish lines, advantageously posted. The ground was unfavorable for the location of artillery on the Confederate side, for, to be effective, the guns would have to move up close to the Federal lines, and that, too, under fire of both infantry and artillery. . . . I said, 'General, it cannot be done with fifty guns and the troops you have near here.' In an instant he said, 'Let us ride back, Colonel.' I felt that I had positively shown a lack of nerve, and with considerable emotion begged that I might be allowed to make the attempt, saying, 'General, you forced me to say what I did unwillingly. If you give the fifty guns to any other artillery officer, I am ruined for life. I promise you I will fight the guns to the last extremity, if you will only let me command them.' Jackson said, 'It is all right, Colonel. Everybody knows you are a brave officer and will fight the guns well. Go to General Lee, and tell him what has occurred since you reported to me. Describe our ride to the hill, your examination of the Federal position, and my conversation about your crushing the Federal right with fifty guns, and my forcing you to give your opinion." Stephen Lee did so. The story behind this event is that Lee had ordered Jackson on the evening of September 17 to turn the enemy's right, and Jackson said it couldn't be done. Lee reiterated the order on the 18th and told Jackson to take fifty guns and crush the Federal right. Jackson said that if an artillerist in whom General Lee had confidence would say the Federal right could be crushed with fifty guns, he would make the attempt. Lee chose Stephen Lee, and Jackson forced him to give his opinion. Stephen Lee's negative opinion agreed with Jackson's view, and that ended the matter. See Henderson, vol. 2, 264–67. (Henderson's source was a letter from Stephen Lee to Henderson.)

9. Johnson and Buel, vol. 2, 679–80.

Chapter 9: Fredericksburg

1. Two days later Lincoln overrode another provision of the Constitution, suspending the writ of habeas corpus, which was designed to prevent government au-

thorities from illegally detaining people. Lincoln denied habeas corpus to those persons accused of "discouraging volunteer enlistments, resisting militia drafts or guilty of any disloyal practice, affording aid and comfort to Rebels." The Constitution (Article I, Section 9) permits suspension of the writ of habeas corpus only "in cases of rebellion or invasion" when "the public safety may require it." Roy P. Basler, *The Collected Works of Abraham Lincoln* (New Brunswick, N.J.: Rutgers University Press, 1953–55), vol. 5, 537; John Hope Franklin, *The Emancipation Proclamation* (Garden City, N.Y.: Doubleday, 1963), 61–62, 66–67; Freeman, *R. E. Lee,* vol. 2, 419.

2. Sears, 319–25; Porter Alexander, *Military Memoirs,* 276; Maurice, 157–58, 165; Basler, vol. 5, 442, 508.

3. Porter Alexander, *Military Memoirs,* 277; Palfrey, 130; Allan, *Army of Northern Virginia,* 450–52; Sears, 325.

4. Freeman, *R. E. Lee,* vol. 2, 415–17; Porter Alexander, *Military Memoirs,* 278–81; *War of the Rebellion,* vol. 19, part 2, 625, 629, 640, 656–57, 660, 674, 679, 713, 722. On November 20, 1862, the 1st Corps under Longstreet contained 34,916 men and ninety-nine cannons; the 2nd Corps under Jackson counted 31,692 men and ninety-eight cannons. Stuart's cavalry amounted to 9,146 men and twenty-two guns, and William N. Pendleton's reserve artillery had thirty-six guns and 718 men. The army consolidated most of the remaining independent batteries into artillery battalions. Jackson kept his artillery attached to the separate divisions, while Longstreet allowed some artillery to remain with divisions but formed as corps reserve artillery the nine-gun battalion of the Washington Artillery of New Orleans and the twenty-six-gun battalion of Colonel E. Porter Alexander (formerly under Stephen D. Lee, promoted to brigadier general).

5. *War of the Rebellion,* vol. 19, part 2, 633–34; 698–99; Freeman, *R. E. Lee,* vol. 2, 417–18; Porter Alexander, *Military Memoirs,* 278. Jackson's and Longstreet's promotions were announced on November 6, 1862. The army's cavalry also was reorganized into four brigades, and Lee's son Rooney was promoted to brigadier general and named to command one of the brigades.

6. Allan, *Army of Northern Virginia,* 456–57; Porter Alexander, *Military Memoirs,* 281–82; Sears, 339–44; Freeman, *R. E. Lee,* vol. 2, 428. Burnside was an 1847 graduate of West Point, resigned from the army in 1853 to manufacture a new breech-loading rifle of his design, failed, got a railroad job through the efforts of George McClellan, commanded a brigade at First Manassas, captured Roanoke Island and New Bern in North Carolina in early 1862, and gained command of the 9th Corps, but at Antietam showed almost no initiative or imagination.

7. *War of the Rebellion,* vol. 19, part 2, 546, 552; Allan, *Army of Northern Virginia,* 459.

8. At this stage, the Union had far more artillery than the Confederacy and also a far higher proportion of long-range rifled pieces. In the Army of the Potomac

the ratio of rifled to smoothbore cannons was seven to three, in the Army of Northern Virginia two to three. The maximum *effective* range of rifles was about 2,500 yards and of smoothbores 1,500 yards, though many could not reach that distance. Cannons fired solid shot from 350 yards out, shrapnel from 500 to 1,000 yards, canister from 500 yards in. The Federals had far more of the best rifled pieces used by both sides: twenty-pounder Parrott guns with a maximum range of 4,500 yards; ten-pounder Parrott guns with a maximum range of 6,200 yards; and three-inch ordnance guns, range 4,180 yards. The Army of the Potomac also used a twelve-pounder smoothbore with a range of 1,660 yards, and both sides used the light twelve-pounder Napoleon cannon with a range of 1,300 yards. The Napoleon was increasingly employed up close to defenders' lines, where it was loaded with canister and became a powerful weapon to assist riflemen in fending off attacks. In these situations, the Napoleons generally fired at ranges of 300 to 500 yards. A large portion of Confederate smoothbores were six-pounder guns, range 1,525 yards, and twelve-pounder howitzers, range 1,070 yards, calibers discarded by the Federals as too light. Confederate ammunition also was defective, Federal largely good. Nearly all cannons on both sides were muzzle loaders. The Federals also had a few 4.5-inch "ordnance" guns and the Confederates a few English Blakely and Whitworth rifles. "Pounder" referred to the weight of the projectile or charge fired, "inch" to the diameter of the bore. See Porter Alexander, *Military Memoirs,* 280; Bigelow, 22–23, 27–28; Ian V. Hogg, *Illustrated Encyclopedia of Artillery* (Secaucus, N.J.: Chartwell Books, 1988), 92, 194, 248–49; Diagram Group, *Weapons, an International Encyclopedia* (New York: St. Martin's Press, 1990), 171, 174–75, 179.

9. Dabney, 595; James Longstreet in Johnson and Buel, vol. 3, 71–72; Henderson, vol. 2, 304; Porter Alexander, *Military Memoirs,* 287–88. An editor's footnote in Johnson and Buel (cited above) showed that Lee, in a dispatch to Richmond authorities on the second day after the battle of Fredericksburg, wrote that if the enemy crossed at Port Royal, he thought it best to withdraw to "the Annas and give battle, than on banks of the Rappahannock." Lee said he had wanted to do this at first, but didn't want to open any more of the country to depredation than necessary, and also wanted to collect forage and provisions in the Rappahannock valley.

10. Dabney, 595–96; Johnson and Buel, vol. 3, 72; Porter Alexander, *Military Memoirs,* 288; Freeman, *R. E. Lee,* vol. 2, 431, 439; *War of the Rebellion,* vol. 21, 1029; Henderson, vol. 2, 301–6.

11. The general trend of the Rappahannock River is eastward. However, at Fredericksburg the river turns in a southeasterly direction. In order to simplify the account, the river is regarded as running due eastward, and locations are given as north or south of the river.

12. Johnson and Buel, vol. 3, 130.

13. Palfrey, 152–53, 160; Porter Alexander, *Military Memoirs,* 294; Freeman,

R. E. Lee, vol. 2, 457–60; *War of the Rebellion,* vol. 21, 70, 90–91; Vorin E. Whan Jr., *Fiasco at Fredericksburg* (State College: Pennsylvania State University Press, 1961), 54–55, 62.

14. Porter Alexander, *Military Memoirs,* 302.

15. Ibid., 313. Johnson and Buel, vol. 3, 145, 147, have slightly different figures.

Chapter 10: Chancellorsville

1. Johnson and Buel, vol. 3, 84.

2. Freeman, *R. E. Lee,* vol. 2, 477.

3. Compounding problems for the Confederate army was the incompetence of the commissary general, Colonel Lucius B. Northrop. Northrop seriously deprived the soldiers of supplies over the winter but refused to improve his system. Nevertheless, he remained in office because he was supported by President Davis. Only the efforts of soldiers' families, who sent food from home, prevented severe repercussions.

4. Thomas, 272.

5. Freeman, *R. E. Lee,* vol. 2, 478–79, 499–501; Bigelow, 114–19; *War of the Rebellion,* vol. 21, 1096–97. One reason for Lee's lack of resistance to President Davis was a serious illness that struck him in late March 1863. He called it a severe cold, but the symptoms included an elevated pulse rate and pain in his chest, back, and arms. After two weeks in bed, Lee still felt weak and unsteady, and he never completely recovered. His ailment may have been angina pectoris, caused by atherosclerosis, or gradual constriction of the blood flow in his arteries. The attack was the signal for the onset of cardiovascular problems that resulted in his death from a stroke in 1870.

6. He said of the battle of Fredericksburg that he went in and fought his troops "until he thought he had lost as many men as he was ordered [by Burnside] to lose." See Theodore A. Dodge, *The Campaign of Chancellorsville* (Boston: James R. Osgood and Company, 1881), 13.

7. After the "mud march" Burnside returned to camp at Falmouth blaming his generals for the failure, especially Joseph Hooker. He relieved Generals William B. Franklin and William Farrar Smith, plus six division or brigade commanders. Burnside told Lincoln he could not continue in his job unless the president approved these actions. Faced with an ultimatum, Lincoln quashed Burnside's order and removed him from command. Many Union officers wanted McClellan reinstated, but Lincoln knew this would be tantamount to military dictatorship. Lincoln dismissed Franklin for his lack of effort after two of his divisions had been repulsed by Jackson at Fredericksburg. Lincoln didn't consider Edwin V. Sumner because he was growing old and feeble and was also a partisan of McClellan.

8. Bigelow, 8–10; Johnson and Buel, vol. 3, 216–17, 239; Porter Alexander, *Military Memoirs,* 373; Freeman, *Lee's Lieutenants,* vol. 2, 429; Freeman, *R. E. Lee,* vol. 2, 484.

9. Henderson, vol. 2, 344–45.

10. The Orange Plank Road and Orange Turnpike ran together out of Fredericksburg for about five miles to a point just east of Zoan Church, about four and a half miles east of Chancellorsville. At Zoan the Plank Road arched southward but rejoined the turnpike at Chancellorsville. The two roads ran together for about two miles to Dowdall's Tavern (also the location of Wilderness Church), where the Plank Road again diverged south and pursued a more southerly course to Orange Court House. Present-day Virginia Route 3 pursues the course of the old turnpike to Wilderness Tavern (two and a half miles west of Dowdall's Tavern), where it becomes Virginia Route 20 and turns southwest to Orange. Virginia Route 610 traces the southerly route of the Plank Road from Zoan to Chancellorsville, and Virginia Route 621 the southerly course of the Plank Road west from Dowdall's Tavern. It rejoins Virginia Route 20 at Verdiersville.

11. Doubleday, 4–5.

12. Federal cavalry were largely armed with the Sharps single-shot breechloading carbine. It gave a rate of fire higher than the infantry rifle musket, but its bullet had an effective range of only about 175 yards, and followed an extreme parabolic trajectory that made it difficult to control. Since the infantry rifle had an effective range of 400 yards, cavalry at best could deflect an infantry attack temporarily, not hold a position, especially since cavalry usually brought along only a few cannons. Also, since the cavalry strike was in the enemy rear, the horsemen would soon run out of supplies, especially ammunition. Their presence in the rear could only be sustained provided the main Federal army moved quickly to their support.

13. Porter Alexander, *Fighting for the Confederacy,* 195.

14. Accounts also listed this church as Zion and Zoar. The church was on the turnpike about a mile and a quarter west of the junction of the turnpike and the Orange Plank Road and was immediately west of the entrenchments Anderson and McLaws were constructing on the morning of May 1, 1863. This position also was known as Tabernacle Church, named for the house of worship on the Mine Road, three-quarters of a mile southwest of the turnpike–Plank Road junction. The Mine Road cut across the Plank Road and joined the turnpike a third of a mile east of Zoan Church. The intersections of all three of these roads collectively were known as Zoan Church or Tabernacle Church. They are carried here as Zoan Church.

15. Henderson, vol. 2, 417; Bigelow, 223.

16. During the morning of May 1, 1863, Hooker received reports from three captive balloons that he had hoisted over the terrain and also from signal observation posts he had set up on heights. The balloons and signal officers reported two

corps, about 15,000 men, were moving toward Chancellorsville (this was Jackson's force). The night before, deserters had told Federal officers that Hood's division had rejoined Lee, coming from Suffolk. This was false, and could have been confirmed by telegraphing Fort Monroe. Hooker had confidently expected Lee to retreat without a battle. Finding instead that Lee had turned on him like a lion, Hooker, in Porter Alexander's words, "lost his nerve and wished himself back on the line he had taken around Chancellorsville, where he would enjoy the great advantage of acting upon the defensive." Hooker's distraction can be shown in a strange wire he sent at 2:00 P.M. May 1 to his chief of staff, Major General Daniel Butterfield, at Falmouth: "From character of information have suspended attack. The enemy may attack me—I will try it. Tell Sedgwick to keep a sharp lookout and attack if he can succeed." See Porter Alexander, *Military Memoirs,* 327; Bigelow, 250–51; Freeman, *Lee's Lieutenants,* vol. 2, 647; *War of the Rebellion,* vol. 25, part 2, 326, 328.

17. Johnson and Buel, vol. 2, 161; Bigelow, 259.

18. Porter Alexander, *Fighting for the Confederacy,* 216.

19. *Southern Historical Society Papers,* vol. 25, 110; Bigelow, 340n. Robert Lewis Dabney writes that Jackson, speaking afterward, "said if he had had an hour more of daylight or had not been wounded, he should have occupied outlets toward Ely's and United States Fords, as well as those on the west. . . . If he had been able to do so dispersion or capture of Hooker's army would have been certain." See Dabney, 699–700.

20. Henderson, vol. 2, 431–32. (Henderson's source was a personal letter from Jedediah Hotchkiss.)

21. Hawkins Farm was also given in some reports as the Taylor house. It was just northwest of the point where the Plank Road turned off to the southwest from the turnpike. Nearly at this junction were also the Wilderness Church and Dowdall's Tavern, also carried in some reports as Melzi Chancellor's house.

22. Johnson and Buel, vol. 3, 198.

23. Although Rodes's and Colston's divisions had become mingled, Jackson galloped into the disordered ranks and directed the men to press the pursuit despite the disorganization. See Henderson, vol. 2, 445. Thus Rodes disobeyed his chief's explicit instructions. R. L. Dabney is incorrect when he says that Jackson determined to relieve the front line and replace it with A. P. Hill's division. See Dabney, 683. Bigelow, p. 337, writes that Jackson had issued direct orders that there be no pause in the advance and was disappointed by Rodes's halt. Neither Rodes nor Colston asserted or implied that the troops were exhausted or incapable of further effort. See *War of the Rebellion,* vol. 25, part 1, 941, 1004–5; Colston, in Johnson and Buel, vol. 3, 233; Henderson, vol. 2, 448.

24. Henderson (vol. 2, 445) points out that the Southerners were only one and a half miles away from the Federal center and completely in the rear of the Union

entrenchments. "And the White House or Bullock Road, only half a mile in front," he wrote, "led directly to Hooker's line of retreat by the United States Ford. Until that road was in his possession Jackson was determined to call no halt."

25. Some of these artillery pieces came from the 11th Corps; Hooker's staff had stopped the men as they were fleeing. Captain T. W. Osborn, chief of Berry's artillery, wrote: "As we passed General Hooker's headquarters, a scene burst upon us which God grant may never again be seen in the Federal army of the United States. The 11th Corps had been routed and were fleeing to the river like scared sheep. The men and artillery filled the road, its sides and the skirts of the field. . . . Aghast and terror-stricken, heads bare and panting for breath, they pleaded like infants at the mother's breast that we should let them pass to the rear unhindered." *War of the Rebellion,* vol. 25, part 1, 483; Bigelow, 311.

26. Henderson, vol. 2, 449; Freeman, *Lee's Lieutenants,* vol. 2, 564–65; *Southern Historical Society Papers,* vol. 6, 267.

27. Bigelow, 338.

28. Porter Alexander (*Fighting for the Confederacy,* 212–13) wrote: "I have always felt surprise that the enemy retained Sedgwick as a corps commander after that day, for he seems to me to have wasted great opportunities, and come about as near to doing nothing with 30,000 men as it was easily possible to do. He gave his men a rest after taking Marye's Hill, and when he did advance it was so cautiously done that Wilcox's Alabama brigade alone delayed him until about 5 o'clock in getting to Salem Church—about four miles only from Fredericksburg." This gave time for McLaws's division to come up, and, with Wilcox, drive Sedgwick back. After this Sedgwick went into camp. The next day as well Sedgwick did practically nothing, finally retreating over Banks Ford.

29. Dabney, 723; Henderson, vol. 2, 470; Bigelow, 439. For a full account of Jackson's death and burial, see Dabney, 707–35.

30. Johnson and Buel, vol. 3, 202.

Chapter 11: Gettysburg

1. Dodge, 161.

2. Jackson's approach did, of course, differ from Lee's. Jackson's overall aim was to strike at Northern industry, railways, and so forth—"to force the people of the North to understand what it will cost them to hold the South in the Union at the bayonet's point"—whereas Lee wanted to strike at the Federal army directly. But Jackson's plan necessarily would have included taking on, and defeating, a Union force on Northern ground. The Rebel army could never have carried out a campaign like the one Jackson advocated without having met and defeated the Union army. His program would have forced the Union army to challenge him; Lincoln especially would have required it. Recall that in the original proposal Jack-

son made in the fall of 1861, he said the Confederates could "beat McClellan's army if it came out against us in the open country." In 1861 Jackson was sure the Confederates could win because the Union army at that time was composed of recruits and thus could not stand up in a fight. In 1862, when he proposed the same strategy, Jackson again believed he could beat a Union army in an open challenge; the maneuvering skill he demonstrated in the Shenandoah Valley campaign probably accounted for his confidence. There is the possibility, however, that he was already moving to his theory of forcing the Union army to attack the Rebel army, not the other way around. Though Jackson's theory did not appear in full bloom until August 1862, it's likely that the concept had been maturing in his mind beforehand. In any event, he expected to encounter the Union army on Northern ground and defeat it.

3. Woodworth, 228–32; Freeman, *R. E. Lee,* vol. 3, 19.

4. Freeman, *R. E. Lee,* vol. 3, 34–36.

5. The reorganization of the army's artillery was completed about this time. Each corps had five battalions, averaging sixteen guns each. One battalion was assigned to each division and two battalions to the artillery reserve. All could be assembled for cannonades. General William N. Pendleton remained the chief of the army artillery, but his general artillery reserve was broken up and never reestablished. This structure subsequently was adopted by Prussia, Austria, France, and Britain, and was the genesis of the division and corps artillery organization that became standard in the U.S. Army. See Porter Alexander, *Military Memoirs,* 367, 370.

6. Johnson and Buel, vol. 3, 246–47.

7. Porter Alexander, *Fighting for the Confederacy,* 230.

8. Doubleday, 83.

9. *War of the Rebellion,* vol. 27, part 1, 31, 34–35.

10. Today U.S. Routes 17 and 50 go through Ashby's Gap, and Virginia Route 7 passes through Snicker's Gap.

11. On June 23, 1863, Lee proposed to President Davis that General Pierre Beauregard move some troops to Culpeper and threaten Washington. This would force Hooker to detach part of his army to contain the threat and give Lee superior forces in Pennsylvania. But Davis refused to call up Beauregard, and the strategic opportunity was lost.

12. Jenkins exacted heavy contributions of horses, cattle, grain, and other goods, paying in Confederate money. If locals refused Southern currency, officers issued receipts setting forth the name of the owner and the fair market value of the products taken. While there, locals stole some of Jenkins's horses. He sternly demanded compensation. The Chambersburg city fathers paid, in Confederate money. Jenkins took the bills with a smile, appreciating the joke. See Doubleday, 96. General Lee had laid down stringent rules for the conduct of Confederate

troops in the North. His General Orders No. 73 stated: "It must be remembered that we make war only upon armed men, and that we cannot take vengeance for the wrongs our people have suffered without lowering ourselves in the eyes of all whose abhorrence has been excited by the atrocities of our enemies. . . . The commanding general therefore earnestly exhorts the troops to abstain . . . from unnecessary or wanton injury to private property." Although there were some violations of the order, no charges of rape and few of plundering were leveled. See Freeman, *R. E. Lee,* vol. 3, 56–57; Douglas, 245.

13. *War of the Rebellion,* vol. 27, part 3, 931–32; Freeman, *R. E. Lee,* vol. 3, 51; *Southern Historical Society Papers,* vol. 4, 99.

14. Lee's official report on the Gettysburg campaign reads: "In the absence of the cavalry, it was impossible to ascertain his [the enemy's] intentions; but to deter him from marching farther west, and intercepting our communications with Virginia, it was determined to concentrate our army east of the mountains." See *War of the Rebellion,* vol. 27, part 2, 316.

15. Ibid., 317.

16. It will occur to readers that this was precisely how Lee selected Sharpsburg and Antietam as a battle site in September 1862. Lee chose both Sharpsburg and Gettysburg because his troops could reach them from several directions. But, as with Gettysburg, he knew nothing about Sharpsburg's suitability for defense. In the event, the selection was disastrous, because Lee's misjudgment about where the Potomac River ran westward from the town eliminated any possibility of a move around McClellan's flank.

17. The Union right wing consisted of the 2nd, 5th, 6th, and 12th Corps under Henry W. Slocum, who also commanded the 12th Corps.

18. *Southern Historical Society Papers,* vol. 4, 157; Freeman, *R. E. Lee,* vol. 3, 65; *War of the Rebellion,* vol. 27, part 2, 317, 607, 637.

19. Porter Alexander, *Military Memoirs,* 381. Alexander wrote: "Hill's movement to Gettysburg was made on his own motion, and with knowledge that he would find the enemy's cavalry in possession. Ewell was informed of it. Lee's orders were to avoid bringing on an action. Like Stuart's raid, Hill's venture is another illustration of an important event allowed to happen without supervision. Lee's first intimation of danger of collision was his hearing Hill's guns at Gettysburg." See also *War of the Rebellion,* vol. 27, part 2, 317, 444.

20. Longstreet, 357; Freeman, *R. E. Lee,* vol. 3, 67.

21. Howard's performance in opposing Ewell illustrates one of the perils of military operations. A commander must rely on subordinates to carry out orders properly and intelligently. When they do not, their local failure can spell victory or defeat for the whole army. General Doubleday emphasized to Howard that his corps should be placed to stop a Confederate advance from Oak Ridge. Whether Howard made this plain to General Schimmelpfennig is not known. In any event,

when Schimmelpfennig lined up his division to the east of Oak Ridge, leaving Doubleday's flank wide open, Howard did nothing to correct the mistake. Likewise, Barlow deployed his division facing north, with no adequate flank protection on his east. Howard's corps, then, was ripe to be enveloped on both sides. Not only was it unable to provide protection for Doubleday's corps on the west, but it collapsed and fled in chaos the moment Ewell's troops struck it on its open eastern flank.

22. *Southern Historical Society Papers,* vol. 4, 458; Freeman, *R. E. Lee,* vol. 3, 70–71; Freeman, *Lee's Lieutenants,* vol. 3, 86–87. In his official report, Lee wrote that he was seeking "the most favorable point of attack," but he took no responsibility for ordering Heth and Pender to attack. The decision, though, was plainly his. No other officer had the authority to order a general attack. Lee put the onus on Hill, writing that Heth's troops, at the start of the engagement, "drove in the advance of the enemy very gallantly, but subsequently encountered largely superior numbers, and were compelled to retire with loss. . . . General Heth then prepared for action, and as soon as Pender arrived to support him, was ordered by General Hill to advance." See *War of the Rebellion,* vol. 27, part 2, 307–8, 317, 638. A. P. Hill in his official report lapsed into the passive voice to avoid taking responsibility himself or blaming Lee. He wrote: "About 2:30 o'clock, the right wing of Ewell's corps made its appearance on my left, and thus formed a right angle with my line. Pender's division was then ordered forward, [Edward L.] Thomas's brigade being retained in reserve, and the rout of the enemy was complete, [Abner] Perrin's [South Carolina] brigade taking position after position of the enemy, and driving him through the town of Gettysburg." See ibid., 607.

23. Fremantle, 203. Hill, it turned out, was not up to the responsibility of corps command, and this might have been the first sign of his inadequacy. He later was to perform poorly in the Wilderness as well.

24. Porter Alexander, *Fighting for the Confederacy,* 232. Alexander's description of the Federal position was most graphic: "It was in the shape of a fishhook. A straight shank, some two and a half miles long, reached to Big and Little Round Top mountains, forming a very strong left flank, with superb view and command over all the open ground in front, and unlimited positions for the Federals' strongest arm, their fine artillery. Opposite the town the hook gradually bends around [Culp's Hill], and then runs off the point some 2,000 yards long, in a rocky ridge overlooking a creek [Rock Creek], very steep in front in many places, and masked by woods from artillery fire of an assailant, but permitting its use at canister range by the Federals. This was, perhaps, the strongest part of their whole position. The whole length of this line was three and a half miles, just right for their force, and stretching us too much for ours. Plenty of stone fences and wood permitted rapid arrangement of breastworks everywhere."

25. Johnson and Buel, vol. 3, 339–40. Porter Alexander (*Military Memoirs,* 386–87) recounts the exchange as follows. Lee: "If he [the enemy] is there tomorrow,

I shall attack him." Longstreet: "If he is there tomorrow it will be because he wants you to attack him."

26. Jean Colin, *The Transformations of War* (London: Hugh Rees, 1912), 279–89. Napoléon wrote: "A well-established maxim of war is not to do anything which your enemy wishes—and for the single reason that he does so wish. You should therefore avoid a field of battle which he has reconnoitered and studied. You should be still more careful to avoid one which he has fortified and where he has entrenched himself. A corollary of this principle is, never to attack in front a position which admits of being turned." See Major Thomas R. Phillips, ed., *Roots of Strategy* (Harrisburg, Pennsylvania: Military Service Publishing Company, 1941), 407–41 (maxim no. 16); Jay Luvaas, *Napoleon on the Art of War* (New York: Touchstone, 1999), 130.

27. Colonel William Allan, an apologist for Lee, contended (Johnson and Buel, vol. 3, 355) that a movement on Meade's southern flank would have been difficult in the absence of Stuart, and that the movement could not have been made quickly enough to surprise Meade. Neither argument holds water. Albert Jenkins's cavalry brigade of 1,600 men was up and could have scouted out a movement. The key element was not surprise but the move itself. A march on Meade's flank would not have required much more than to set the closest Confederate troops in motion. It could have been done quickly. Meade could not have stopped it. Allan made the odd defense that "if successful it simply would have forced the Federal army back to some position nearer Baltimore and Washington where the issue of battle was still to be tried." Precisely—getting Meade *off* the extremely favorable defensive position of Cemetery Hill and Ridge and forcing the Federals to attack on ground selected by the Confederates was *exactly* what Longstreet was trying to do.

28. Doubleday, 157–58. Doubleday added an intriguing detail in his account: it was an open secret, he wrote, that Meade disapproved of the battleground Hancock had selected. True, Meade had approved Hancock's request to concentrate forces at Gettysburg. But he was a new commander and he had already decided to go on the defensive along Pipe Creek. This was comparatively safe. Moving up to Gettysburg was a dangerous, uncertain move, one made only on Hancock's recommendation. Meade had not seen the site, but he did know that the Federals had been knocked off their positions to the west and had been driven out of the town. When he got there, he disapproved of the battleground, probably because he spotted the same fatal weakness of the position that Longstreet did: all the Confederates had to do to evict him was to move to the south—and then the Rebels would be between his army and Washington, a guarantee that Meade would be tossed out of his command at the first opportunity. Meade was saved only by Lee's decision to attack the Union commander in this position, not maneuver him out of it.

29. Lee wrote twice in his official report (*War of the Rebellion,* vol. 27, part 2, 308, 318) that a battle had become "in a measure unavoidable." In ibid., pp. 307

and 308, Lee wrote that at the end of the first day it was advisable to await the arrival of the rest of the Southern army. "Orders were sent back to hasten their march, and, in the meantime, every effort was made to ascertain the numbers and position of the enemy, and find the most favorable point of attack. [Note that Lee accepted as a given that *he* would attack, not *receive* an attack.] It had not been intended to fight a general battle at such a distance from our base, unless attacked by the enemy, but, finding ourselves unexpectedly confronted by the Federal army, it became a matter of difficulty to withdraw through the mountains with our large trains. At the same time, the country was unfavorable for collecting supplies while in the presence of the enemy's main body, as he was enabled to restrain our foraging parties by occupying the passes of the mountains with regular and militia troops. A battle thus became, in a measure, unavoidable."

Lee's report appeared to be an attempt to justify, after the fact, his decision not only to fight a battle at Gettysburg but to fight an *aggressive* battle there, since he was saying that a battle was more or less forced on him. None of this was true. Only a few lines down in the same report Lee acknowledged that on this day (July 1) he had received word that Stuart and his cavalry had arrived at Carlisle and he had ordered them to move at once to Gettysburg. Therefore, Stuart's cavalry, not to mention Jenkins's and Imboden's horsemen, were available to hold the passes in the mountains to the west, if necessary. Also, the Confederate army, if actually blocked to the west and south, could simply have moved north, where plenty of food was available. He then could have turned toward Philadelphia. If Meade was unwilling to attack, there was nothing to stop Lee from marching away.

In any event, the idea that the Southern army had to attack because otherwise it would run out of food was absolutely false. Porter Alexander remarked (*Fighting for the Confederacy,* 233): "When it is remembered that we stayed for three days longer on the very ground [of Gettysburg], two of them days of desperate battle, ending in the discouragement of a bloody repulse, and then successfully withdrew all our trains and most of the wounded through the mountains; and, finding the Potomac too high to ford, protected them all and foraged successfully for over a week in a very restricted territory along the river, until we could build a bridge, it does not seem improbable that we could have faced Meade safely on the 2nd [of July] at Gettysburg without assaulting him in his wonderfully strong position."

30. Porter Alexander, *Fighting for the Confederacy,* 233–34.

31. In the afternoon on July 2 Meade focused on the fact that Sickles had moved entirely off Cemetery Ridge and had not occupied Little Round Top. Livid, Meade rode over to the Peach Orchard and, realizing the line could not be moved back at such a late hour, ordered George Sykes's 5th Corps from reserve to assist Sickles, provided him artillery from the army reserve, and told him he could call on the 2nd Corps for help if needed. "You cannot hold this position," he told Sickles,

"but the enemy will not let you get away without a fight, and it may as well begin now as at any time." Meade censured Sickles's decision afterward, setting off a dispute that raged for decades after the war. See Pfanz, *Second Day,* 140–44.

32. Johnson and Buel, vol. 3, 320–21.

33. Hood, 59; Freeman, *Lee's Lieutenants,* vol. 3, 117–22. McIvor Law's discovery that Round Top was unoccupied underscores the validity of Longstreet's proposal to Lee to swing around to the south and dislodge Meade from his position. The Confederate army did *not* have to attack the enemy head-on. Even on the afternoon of July 2, a flank attack still would have succeeded.

34. As the attack was about to get under way, Meade sent his chief engineer, General Gouverneur K. Warren, to "the little hill off yonder"—Little Round Top—to make sure it had troops protecting it. Warren found only a Union signal station on the summit and saw at once that the Federals had to hold the hill if they were going to keep Cemetery Ridge. Warren rushed word to Meade, who commenced a frantic effort to feed troops to it. He just barely succeeded.

35. One of the most remarkable aspects of the battle was that Colonel Oates, when he seized Round Top, realized that from this summit far above the surrounding countryside the Confederates could evict the Federals from their entire position if they cut down trees and dragged cannons to the top. But a staff officer reached Oates and ordered him to advance on Little Round Top. General Law had specifically seen that Round Top was the correct *point d'appui,* or base, but he was deeply involved in directing Hood's division and evidently did not know that Oates had seized it. No senior Confederate leader recognized that the prize had already been gained. An advance against Little Round Top or even Cemetery Ridge was not necessary.

36. Union General Doubleday, who was there, wrote: "On this occasion Wright did what Lee failed to accomplish the next day at such a heavy expense of life, *for he pierced our center,* and held it for a short time, and had the movement been properly supported and energetically followed up, it might have been fatal to our army, and would most certainly have resulted in a disastrous retreat. It was but another illustration of the difficulty of successfully converging columns against a central force. Lee's divisions seemed never to strike at the hour appointed. Each came forward separately, and was beaten for lack of support." See Doubleday, 176.

37. Capture of the national capital would turn Lincoln and his ministers into fugitives and destroy the Republican mandate of power. Capture of Baltimore would cut off Washington's food supply, create chaos, and leave Lincoln's administration unable to govern. Capture of Philadelphia would shatter the Northern railroad communication system and make it impossible to continue the war.

38. *War of the Rebellion,* vol. 27, part 2, 320; Freeman, *R. E. Lee,* vol. 3, 112. Despite claims afterward that Lee intended for Longstreet's and Hill's entire corps to advance with Pickett, Lee never contemplated such an advance. See Tucker

337, 343; Johnson and Buel, vol. 3, 356; Long, 294; Walter Taylor, 104, 108. Originally Lee planned to use McLaws's and Hood's divisions, but Longstreet talked him out of it, pointing out that these divisions would have to be withdrawn from the south, giving the Federals an opportunity to pour down from Round Top on the flank and rear of the Confederate army. See Freeman, *R. E. Lee,* vol. 3, 107–8; Longstreet, 386.

39. Johnson and Buel, vol. 3, 342–43; *War of the Rebellion,* vol. 27, part 2, 359; Longstreet, 386; Long, 288; Porter Alexander, *Military Memoirs,* 416.

40. Pickett's division composed a little more than a third of the total. The assaulting columns consisted of forty-two regiments, nineteen from Virginia, fifteen from North Carolina, three each from Tennessee and Mississippi, and two from Alabama. In addition, Wilcox had five Alabama regiments and Lang three small Florida regiments.

41. Lee at first hoped that Ewell on the north could strike Culp's Hill at the same time as Pickett advanced, but the Union 12th Corps assaulted Edward Johnson's division there early on July 3. By 11:00 A.M. Johnson found progress impossible, and withdrew.

42. Porter Alexander, *Fighting for the Confederacy,* 251; *Military Memoirs,* 417–18, 426–28.

43. Johnson and Buel, vol. 3, 364–68.

44. Alexander's message said that "there are still eighteen guns firing from the cemetery." He was actually referring to guns firing from the Copse of Trees, but he had been told incorrectly that the Copse was in the cemetery. Ibid., 344–45.

45. Ibid., 344–45.

46. Ibid., 365; Porter Alexander, *Fighting for the Confederacy,* 261; *Military Memoirs,* 424.

47. Doubleday, 193.

48. Johnson and Buel, vol. 3, 387–90.

49. Doubleday, 195.

50. On the Confederate left, the forces were arrayed as follows, from south to north: Next to Garnett's brigade was Birkett Fry's Alabama-Tennessee brigade, which suffered the same staggering losses as Garnett's force did. To Fry's left or north was Pettigrew's North Carolina brigade, now under James K. Marshall. To Marshall's left was Joseph R. Davis's Mississippi–North Carolina brigade. Directly behind Pettigrew's division was Trimble's division.

51. Johnson and Buel, vol. 3, 347; Fremantle, 213, 215. Commenting on Lee's appearance far forward with the guns, Porter Alexander remarked (Johnson and Buel, vol. 3, 366): "I have since thought it possible that he came, thinking the enemy might follow in pursuit of Pickett, personally to rally stragglers about our guns and make a desperate defense. He had the instincts of a soldier within him as strongly as any man. Looking at Burnside's dense columns swarming through the

fire of our guns toward Marye's Hill at Fredericksburg, he had said: 'It is well war is so terrible or we would grow too fond of it.' No soldier could have looked at Pickett's charge and not burned to be in it. To have a personal part in a close and desperate fight at that moment would, I believe, have been at heart a great pleasure to General Lee, and possibly he was looking for it."

52. Johnson and Buel, vol. 3, 349. In this same citation Longstreet wrote: "During that winter, while I was in east Tennessee, in a letter received from him [Lee] he said, 'If I only had taken your counsel even on the 3rd [of July], and had moved around the Federal left, how different all might have been.' "

53. Rosecrans, with 70,000 men, moved against Bragg, who was resting his 47,000-man army in well-fortified works at Shelbyville. Bragg watched for an attack straight ahead but forgot his flanks. This allowed Union General George H. Thomas to swing all the way around to Decherd, thirty-five miles southeast of Shelbyville, where he cut Bragg's supply line. Thus, with nothing more than maneuver, Rosecrans forced Bragg to withdraw to Chattanooga.

54. Rosecrans crossed the Tennessee River west of Chattanooga. George H. Thomas's corps marched to McLemore's Cove, twenty miles south of Chattanooga, while Alexander McDowell McCook's corps took a more circuitous route and arrived at Alpine, fifteen miles farther south.

55. On September 19, 1863, Bragg sent Leonidas Polk's corps across Chickamauga Creek; the plan was to crush Thomas on Rosecrans's left or eastern flank, then seize Rossville, Georgia, seven miles north, to cut the Union army off from Chattanooga. Thomas held his position through the brutal battle. The next day Polk tried again, and once again Thomas refused to yield. On the right, however, a Union division withdrew because of mistaken orders. The Confederates streamed through the gap and routed the entire Union army except Thomas's corps. Thomas backed up a couple of miles to a good defensive position, Horse Shoe Ridge, and held on grimly, though facing odds of two to one. Bragg tore his columns into shreds by dashing them against "the Rock of Chickamauga."

Chapter 12: Appomattox

1. As a distraction to Joe Johnston in north Georgia, Grant ordered Nathaniel P. Banks to seize the port of Mobile, Alabama, and open an alternative route into Georgia from the extreme south. As a rear threat to Lee, Grant directed Benjamin F. "Beast" Butler to advance westward along the south bank of the James River toward Richmond with about 37,000 men. In the event, neither flanking movement succeeded. A Confederate force of 24,000 under Pierre Beauregard bottled Butler up against the James east of Richmond, while Banks was so slow that his threat against southern Georgia never materialized.

2. Porter Alexander observed: "It is a fact that Johnston had never fought but

one aggressive battle, the battle of Seven Pines [1862], which was phenomenally mismanaged." See Porter Alexander, *Military Memoirs,* 577.

3. Hattaway and Jones, 250, 300, 317, 357; Earl Schenck Miers, *The General Who Marched to Hell* (New York: Dorset, 1990), 218.

4. Either consciously or unconsciously Sherman was emulating the "plan with branches" doctrine advanced in the eighteenth century by the French commander Pierre de Bourcet (1700–1780). This called for a commander to disperse his army into several separate columns that threatened several objectives, forcing the enemy to divide his forces to protect each of them. The commander then could bring one or more of his columns together to strike at one of the targets. Sherman also used this method in his advance on Savannah and in his march through the Carolinas.

5. With Atlanta invested, almost the entire Confederate cavalry force in Georgia, under Joseph Wheeler, commenced long-range raids to break Sherman's rail connection with Chattanooga. His horsemen cut several points between Marietta and Dalton. Sherman discovered, however, that Wheeler did not intend to keep the line broken, which would have forced Sherman to send back a major relief expedition to reopen it. Instead, Wheeler rode into Tennessee, hoping to cut rail lines there and induce Sherman to retreat. This had no hope of success. Men on horseback presented a large and easy target for rifle-firing infantry. To fight infantry, cavalry had to dismount, thereby losing their mobility. Cavalry, consequently, had to adopt a hit-and-run policy, because enemy infantry guarding rail lines could soon surround and destroy any static cavalry force. With Wheeler gone, Sherman repaired the rail line back to Chattanooga, where he had a large stock of food and other goods. Wheeler's proper course would have been to concentrate wholly on breaking repeatedly the railway from Chattanooga to Atlanta. If the horsemen struck one point along the line after another, delivery of food supplies to the army at Atlanta could have been halted.

6. General Sherman described the art of field fortification as follows: "Troops, halting for the night or for a battle, faced the enemy; moved forward to ground with a good outlook to the front; stacked arms; gathered logs, stumps, fence rails, anything which would stop a bullet; piled these to their front, and, digging a ditch behind, threw the dirt forward, and made a parapet which covered their persons as perfectly as a granite wall." See Johnson and Buel, vol. 4, 248. Field fortifications were constructed in varying degrees of perfection, depending on how much time the soldiers had to build them. A trench, or ditch, was rarely wider than three feet or deeper than two feet. A parapet was usually about two and a half feet high, reinforced when possible with logs or other solid objects. Hence the distance from the top of the parapet to the bottom of the ditch was about four and a half feet. Porter Alexander drew a cross section of works in *Fighting for the Confederacy,* 409.

7. William Swinton, *Campaigns of the Army of the Potomac* (New York: 1866), 487.

8. The horrible nature of trench warfare was being thrust upon military leaders. During the period June 3–12, 1864, while the two armies lay facing each other at Cold Harbor, John Gibbon's single Union division reported 280 men killed or wounded by sharpshooter fire. Porter Alexander told of a twelve-pounder Napoleon cannon in Henry Colter Cabell's artillery battalion that, when being tipped to replace wheels that had been shot to pieces, disgorged thirty-seven Minié balls from its barrel, thus showing the intensity of the fire. See Porter Alexander, *Fighting for the Confederacy,* 412.

9. J. F. C. Fuller, *Military History of the Western World* (New York: Funk and Wagnalls, 1954–57; New York: Da Capo, 1987), vol. 3, 15. In 1863, the Confederate government built a line joining Danville, Virginia, and Greensboro, North Carolina. Until then, the main line through central North Carolina went from Greensboro through Raleigh, connected by the Weldon Railroad with Richmond. See Porter Alexander, *Fighting for the Confederacy,* 124; Freeman, *R. E. Lee,* vol. 3, 168; *War of the Rebellion,* vol. 29, part 2, 736.

10. Neither the Tredegar Iron Works at Richmond nor the port of Wilmington, North Carolina, was vital to the Confederacy. The ironworks made cannons, but the Rebels had sufficient artillery and could capture more from the Federals. By moving into the interior, Confederate forces could exploit local food and other resources. The one great need was gunpowder, and the South possessed the world's largest powder factory at Augusta, Georgia. Lee might have followed a strategy similar to that of Frederick the Great of Prussia in the Seven Years' War of 1756–63. Frederick did not have sufficient strength to destroy his enemies, but he won partial victories, kept his own army in existence, and waited for an opportunity that would give him victory. This occurred in 1762, when Peter III became czar of Russia and withdrew from the war.

11. Freeman, *R. E. Lee,* vol. 3, 496n.

12. Lee also refused to consider an alternative to a war of movement—guerrilla warfare. The South was an ideal region to pursue this kind of war. The Confederates were filled with a sense of rightness in their cause and the population was friendly. They could hide, feed, and clothe partisans and spy on Federals. The South was large, with wide rivers, deep forests, mountains impossible to capture, and a poor communications system that would have placed severe limits on the movements of Union armies. Sometime in the spring of 1864 Lee told President Davis that, "With my army in the mountains of Virginia, I could carry on this war for twenty years longer." See J. William Jones, *Life and Letters of Robert Edward Lee, Soldier and Man* (Washington, D.C.: 1906), 295; Freeman, *R. E. Lee,* vol. 3, 496n. A guerrilla strategy implied breaking the Confederate field armies into small units operating out of safe bases in mountains, forests, or swamps, or submerging

like "fish" in the "water" of the civilian population. Guerrillas could strike at enemy railways, convoys, occupied cities, depots, and isolated units, then retreat to their bases or take off their uniforms and return temporarily to civilian pursuits. The Federals would have been required to disperse widely to protect supply lines and bases. Guerrillas could have decided when, where, how, and in what strength to attack, giving them the initiative, and turning the Union army into a passive occupying force that could only respond to the actions of the guerrillas. So long as the South remained defiant, the North never could have won.

13. Grant authorized Sherman's march through the Carolinas not because he understood the strategic advantage of cutting through the remaining heart of the Confederacy and onto the rear of Lee, but because he had learned it would take transports two months to bring Sherman's army to Virginia.

14. Most of these Confederate troops were the remains of Hood's army at Tupelo, Mississippi, who had rushed through Georgia by way of Augusta to reinforce Johnston.

15. Liddell Hart, *Sherman: Soldier, Realist, American,* 356; Sherman, 271.

16. Sherman's remorseless pattern of deliberate *personal* injury to the Southern people sowed seeds of hate that bore bitter fruit. The purpose of war is a more perfect peace. Sherman's legacy was the opposite. The memory of the damage he and his men did was passed from parent to child throughout the South for a century after the war. Sherman's march evoked an enduring folk memory of wanton havoc that embittered the Southern people against the North, the Republican Party, and the national government for generations. This is why the South remained "solid" in voting Democratic for many years.

17. Liddell Hart, *Sherman: Soldier, Realist, American,* 369.

18. Lee continued his fixation on frontal attacks almost to the end. On March 25, he ordered John B. Gordon to seize Fort Stedman in a frontal strike with half his army. Stedman was a strongpoint just east of Petersburg. Despite extreme valor on the part of the Rebel attackers, the attack failed miserably against overwhelming fire. Lee lost nearly 3,000 prisoners and 2,000 men killed and wounded, one-tenth of his remaining army.

Bibliography

Alexander, Bevin. *Lost Victories. The Military Genius of Stonewall Jackson.* New York: Henry Holt and Company, 1992; New York: Hippocrene Books, 2005.

———. *Robert E. Lee's Civil War.* Holbrook, Mass.: Adams Media Corporation, 1998.

Alexander, E. Porter. *Military Memoirs of a Confederate: A Critical Narrative.* New York: Charles Scribner's Sons, 1907; New York: Da Capo Press, 1993.

———. *Fighting for the Confederacy: The Personal Recollections of General Edward Porter Alexander.* Chapel Hill: University of North Carolina Press, 1989.

Allan, William. *The Army of Northern Virginia in 1862.* Boston: Houghton Mifflin and Company, 1892; New York: Da Capo Press, 1995.

———. *History and Campaign of Gen. T. J. (Stonewall) Jackson in the Shenandoah Valley of Virginia.* Philadelphia: J.B. Lippincott, 1880; New York: Da Capo Press, 1995.

Beringer, Richard E., Herman Hattaway, Archer Jones, and William N. Still Jr. *Why the South Lost the Civil War.* Athens: University of Georgia Press, 1986.

Bigelow, John, Jr. *The Campaign of Chancellorsville.* New Haven, Conn.: Yale University Press, 1910.

Bowers, John. *Chickamauga and Chattanooga: The Battles That Doomed the Confederacy.* New York: HarperCollins, 1994.

Carmichael, Peter S., ed. *Audacity Personified: The Generalship of Robert E. Lee.* Baton Rouge: Louisiana State University Press, 2004.

Chambers, Lenoir. *Stonewall Jackson.* 2 vols. New York: William Morrow, 1959.

Clausewitz, Karl von. *On War.* Hammondsworth, England: Penguin Books, 1968; New York: Dorset Press, 1991.

Connelly, Thomas L. *The Marble Man: Robert E. Lee and His Image in American Society.* New York: Alfred A. Knopf, 1977.

Connelly, Thomas L., and Archer Jones. *The Politics of Command: Factions and Ideas in Confederate Strategy.* Baton Rouge: Louisiana State University Press, 1973.

Cooke, John Esten. *Stonewall Jackson: A Military Biography.* New York: D. Appleton and Company, 1876.

Dabney, Robert Lewis. *Life and Campaigns of Lieut.-Gen. Thomas J. (Stonewall) Jack-*

son. New York: Blelock and Company, 1866; Harrisonburg, Va.: Sprinkle Publications, 1983.

Davis, William C. *Battle at Bull Run.* Baton Rouge: Louisiana State University Press, 1977.

Dodge, Theodore Ayrault. *A Bird's-eye View of Our Civil War.* Boston: Houghton, Mifflin and Company, 1883, 1897.

Doubleday, Abner. *Chancellorsville and Gettysburg.* New York: Charles Scribner's Sons, 1882.

Douglas, Henry Kyd. *I Rode with Stonewall.* Chapel Hill: University of North Carolina Press, 1940, 1968.

Fellman, Michael. *The Making of Robert E. Lee.* New York: Random House, 2000.

Freeman, Douglas Southall. *R. E. Lee, a Biography.* 4 vols. New York and London: Charles Scribner's Sons, 1934–35.

———. *Lee's Lieutenants: A Study in Command.* 3 vols. New York: Scribner's, 1942–46.

Fremantle, Arthur James Lyon. *The Fremantle Diary.* Edited by Walter Lord. Boston: Little, Brown and Co., 1954.

Fuller, J. F. C. *The Generalship of Ulysses S. Grant.* New York: Dodd, Mead, 1929; Bloomington: Indiana University Press, 1958.

———. *Grant and Lee: A Study in Personality and Generalship.* London: Eyre & Spottiswoode, 1933; Bloomington: Indiana University Press, 1957.

Gallagher, Gary W., ed. *Lee the Soldier.* Lincoln: University of Nebraska Press, 1996.

———. *The Confederate War: How Popular Will, Nationalism, and Military Strategy Could Not Stave Off Defeat.* Cambridge: Harvard University Press, 1997.

Harsh, Joseph L. *Confederate Tide Rising: Robert E. Lee and the Making of Southern Strategy, 1861–1862.* Kent, Ohio: Kent State University Press, 1998.

———. *Taken at the Flood: Robert E. Lee & Confederate Strategy in the Maryland Campaign of 1862.* Kent, Ohio: Kent State University Press, 1998.

Hattaway, Herman, and Archer Jones. *How the North Won.* Urbana: University of Illinois Press, 1991.

Henderson, Colonel G. F. R. *Stonewall Jackson and the American Civil War.* 2 vols. New York: Longmans, Green and Co., 1898; 1 vol., New York: Longmans, Green and Co., 1936, 1937, 1943, 1949; 2 vols., New York: Konecky & Konecky, 1993.

Henry, Robert Selph. *Nathan Bedford Forrest: First with the Most.* New York: Mallard Press, 1991.

Hood, John B. *Advance and Retreat: Personal Experiences in the United States and Confederate Armies.* New Orleans, 1880; Bloomington: Indiana University Press, 1959.

Humphreys, Andrew A. *The Virginia Campaign, 1864 and 1865.* New York: 1883; reprint, New York: Da Capo Press, 1995.

Johnson, Robert U., and C. C. Buel, eds. *Battles and Leaders of the Civil War.* 4 vols. New York: Century Magazine, 1887–88; reprint, Secaucus, N.J.: Castle, n.d.

Johnston, Joseph E. *A Narrative of Military Operations.* Bloomington: Indiana University Press, 1959; New York: Kraus, 1969.

Jones, Archer. *Civil War Command and Strategy: The Process of Victory and Defeat.* New York: Free Press, 1992.

Lee, Robert E. *Lee's Dispatches, Unpublished Letters of General Robert E. Lee, C.S.A., to Jefferson Davis and the War Department of the Confederate States of America, 1862–65.* Edited by Douglas Southall Freeman. New York: G. P. Putnam's Sons, 1957 (originally published in a limited edition, 1915); Baton Rouge: Louisiana State University Press, 1994.

Liddell Hart, Basil H. *Sherman: Soldier, Realist, American.* New York: Dodd, Mead, 1929.

Long, A. L. *Memoirs of Robert E. Lee.* Charlottesville, Va.: 1886; Secaucus, N.J.: The Blue and Grey Press, 1983.

Longstreet, James. *From Manassas to Appomattox.* Philadelphia: J.B. Lippincott, 1903.

Luvaas, Jay. *The Military Legacy of the Civil War: The European Inheritance.* Chicago: University of Chicago Press, 1959; Lawrence: University Press of Kansas, 1988.

Marshall, Charles. *An Aide-de-Camp of Lee.* Edited by Sir Frederick Maurice. Boston: Little, Brown and Company, 1927.

Matter, William D. *If It Takes All Summer: The Battle of Spotsylvania.* Chapel Hill: University of North Carolina Press, 1988.

Maurice, Major General Sir Frederick. *Robert E. Lee the Soldier.* New York: Houghton Mifflin Co., 1925; New York: Bonanza Books, n.d.

McKenzie, John D. *Uncertain Glory: Lee's Generalship Re-examined.* New York: Hippocrene Books, 1997.

Miers, Earl Schenck, and Richard A. Brown. *Gettysburg.* New Brunswick, N.J.: Rutgers University Press, 1948; Armonk, N.Y.: M. E. Sharpe, 1996.

Nolan, Alan T. *Lee Considered: General Robert E. Lee and Civil War History.* Chapel Hill: University of North Carolina Press, 1991.

Nosworthy, Brent. *The Bloody Crucible of Courage: Fighting Methods and Combat Experience of the Civil War.* New York: Carroll and Graf, 2003.

Palfrey, Francis Winthrop. *The Antietam and Fredericksburg.* New York: Charles Scribner's Sons, 1882.

Pfanz, Harry W. *Gettysburg—the Second Day.* Chapel Hill: University of North Carolina Press, 1987.

———. *Gettysburg—Culp's Hill and Cemetery Hill.* Chapel Hill: University of North Carolina Press, 1993.

Power, J. Tracy. *Lee's Miserables: Life in the Army of Northern Virginia from the Wilderness to Appomattox.* Chapel Hill: University of North Carolina Press, 1998.

Rhea, Gordon C. *The Battle of the Wilderness, May 5–6, 1864.* Baton Rouge: Louisiana State University Press, 1994.

Robertson, James I., Jr. *Stonewall Jackson: The Man, the Soldier, the Legend.* New York: Macmillan Publishing USA, 1997.

Ropes, John C. *The Army Under Pope.* New York: Charles Scribner's Sons, 1881.

Scaff, Morris. *The Battle of the Wilderness.* Boston: Houghton Mifflin Co., 1910.

Sears, Stephen D. *Landscape Turned Red: The Battle of Antietam.* New York: Ticknor and Fields, 1983.

———. *To the Gates of Richmond: The Peninsula Campaign.* New York: Ticknor & Fields, 1992.

Selby, John. *Stonewall Jackson as Military Commander.* London: B.T. Batsford Ltd., n.d.; Princeton, N.J.: D. Van Nostrand Co., 1968.

Sherman, William Tecumseh. *The Memoirs of General William T. Sherman.* Bloomington: University of Indiana Press, 1977.

Southern Historical Society Papers. 50 vols. Richmond, Va.: 1876–1953.

Taylor, Richard. *Destruction and Reconstruction. Personal Experiences in the Late War.* New York: D. Appleton and Co., 1879; Nashville, Tenn.: J.S. Sanders and Co., 1998.

Taylor, Walter D. *Four Years with General Lee.* New York: D. Appleton, 1877; New York: Bonanza Books, 1962.

Thomas, Emory M. *Robert E. Lee: A Biography.* New York: W. W. Norton, 1995.

Tucker, Glenn. *High Tide at Gettysburg.* Boston: Bobbs-Merrill Co., 1958; New York: Smithmark Publishers, 1995.

Vandiver, F. E. *The Mighty Stonewall.* New York: McGraw-Hill Book Co., 1957.

The War of the Rebellion: A Compilation of the Official Records of the Union and Confederate Armies. 128 parts in 70 vols., and atlas. Washington, D.C.: Government Printing Office, 1880–1901. Available online at Cornell University Library Digital Library Server, Making of America, Collection, http://cdl.library.cornell.edu.

Weigley, Russell F. *The American Way of War: A History of United States Military Strategy and Policy.* New York: Macmillan, 1973.

Wiley, Bell Irvin. *The Life of Johnny Reb.* Baton Rouge: Louisiana State University Press, 1952, 1978, 1995.

———. *The Life of Billy Yank.* Baton Rouge: Louisiana State University Press, 1952, 1978, 1995.

Wise, Jennings Cropper. *The Long Arm of Lee: The History of the Artillery of the Army of Northern Virginia.* Lynchburg, Va.: J. P. Bell Co., 1915; Oxford University Press, New York, 1959.

Wolseley, Field Marshal Viscount. *The American Civil War: An English View.* Charlottesville: University Press of Virginia, 1964.

Woodworth, Steven E. *Davis and Lee at War.* Lawrence: University Press of Kansas, 1995.

Acknowledgments

I WISH TO THANK most sincerely my editor, Jed Donahue, senior editor at Crown Publishing Group, for his tremendous skill in directing my writing into paths of order and clarity and for his unerring ability to see the goal we were trying to reach and leading me to it. He is a gifted editor, and his contributions have greatly improved this book.

I also want to thank Lauren Dong for her splendid work in designing this book and my old friend Jeffrey L. Ward for his beautiful maps that reveal, with marvelous lucidity, the essential elements of military campaigns and battles.

Finally, I want to acknowledge my ongoing debt to and deep appreciation for my agent, Agnes Birnbaum, for her dazzling capacity to see the real world and for her insistence that I stay focused on it. I am deeply grateful for her friendship, her flawless counsel, and her faith in me.

Index

Peach Tree Creek, battle of, 254

Pelham, John, 181, 187

Pemberton, John Clifford, 209, 210, 246–47, 274*n*

Pender, William Dorsey, 110, 137, 155
at Gettysburg, 221, 223–25, 227, 238, 307*n*

Pendleton, Alexander S., 66, 284*n*

Pendleton, William N., 194, 299*n,* 305*n*

Perrin, Abner, 307*n*

Perryville, battle of, 247

Peter III, Czar of Russia, 314*n*

Petersburg campaign, 42, 258–59, 263, 265

Pettigrew, Johnston, 221, 22, 223, 227–28, 238, 240, 241, 244, 245, 311*n*

Pickett, George E., 6, 185, 210, 220, 238–46, 265, 275*n,* 310*n*–12*n*

Philadelphia, 4, 5, 142, 210, 218, 220, 221, 222, 237, 309*n,* 310*n*

Pierce, Franklin, 33

Pinkerton, Allan, 45, 275*n,* 283*n*

Pipe Creek, 222, 226, 227, 228

Pittman, Samuel F., 151

"plan with branches" doctrine, 254, 313*n*

Pleasonton, Alfred, 147, 192–93, 201, 204, 205, 212, 213, 222

Poague, William T., 256

Polk, Leonidas, 250, 312*n*

Pope, John, 146, 155, 187
Lincoln's dismissal of, 139
mission of, 101, 121, 285*n*–86*n*
personality of, 103
reputation of, 128, 286*n*

Pope, John, campaign against, 98–140, *105,* 174, 176, 198
artillery in, 108, 109, 119
casualties of, 109, 110, 111, 119
cavalry in, 108, 110, 113, 115, 118
Cedar Mountain in, 107–11, 287*n*
forcing enemy to attack in, 100–101, 123, 285*n,* 290*n*
Jackson's deceptive maneuver in, 115–20, 122–23
Jackson's plans in, 106–7, 111–14, 123, 287*n,* 288*n*–89*n*
Lee's delays in, 112–14
Lee's maneuvering in, 98–99, 102, 113, 114–15, 117, 119, 120, 290*n*
missed opportunities in, 111–14
Pope's dispatch book captured in, 115

railroad supply lines severed in, 112, 113, 115–20, 121, 123, 288*n*–90*n*
supplies captured in, 118, 118, 121
telegraph wires cut in, 118, 289*n*–90*n*
troop strength in, 102, 104, 106, 112, 117, 286*n,* 287*n,* 288*n*
see also Second Manassas, battle of,

Porter, Andrew, 20–21

Porter, Fitz John, 117, 288*n*
at Antietam, 163, 166, 171
at Second Manassas, 126, 128, 130, 131, 133, 135, 136, 291*n*
in Seven Days battles, 78–84, 86–87, 88, 90, 91, 95, 98, 280*n,* 281*n*–83*n,* 284*n,* 285*n*

Port Republic, battle of, 72–73, 74–75, 76, 279*n*

Posey, Carnot, 199–200

Potomac River, 12, 47, 49, 60, 66, 77, 146, 155, 156, 161, 173, 275*n*

Prospect Hill, 176, 179–81, 192, 194, 207

Pryor, Roger A., 165–66

railroads, 48, 54, 62, 70, 209, 249, 250, 252, 262, 314*n*
Confederate lateral, 35, 271*n*–72*n*
as military transport, 13–14, 23, 58, 62–63, 76, 106, 169
severing communications with, 4, 31–32, 99, 112, 113, 115–20, 123, 141, 169, 172, 185, 253–54, 259, 265, 288*n*–90*n,* 293*n,* 310*n,* 313*n*
as Sherman's supply line, 250, 253–54, 260, 313*n*

Ramseur, S. Dodson, 202, 203

Randolf, George W., 53, 294*n*

Ransom, Robert, Jr., 185

Reno, Jesse L., 117, 126, 138, 153, 154, 287*n*

Reynolds, John F., 195, 206, 288*n,* 291*n*
at Gettysburg, 222, 223, 225
at Second Manassas, 124, 125, 126, 128, 136, 138

Rice, Edmund, 242, 243–45

Richardson, Israel B., 16, 17–18, 165, 166

Richmond, 8, 9, 10, 12, 22, 68, 99, 209, 257, 262, 271*n*
defense of, 35, 53–54, 57, 60, 70, 76–77, 258
as rail hub, 48

About the Author

BEVIN ALEXANDER is the author of nine books of military history, including *How Hitler Could Have Won World War II, How Wars Are Won, How America Got It Right,* and *Lost Victories,* which was named by the *Civil War Book Review* as one of the seventeen books that have most transformed Civil War scholarship. His battle studies of the Korean War, written during his decorated service as a combat historian, are stored in the National Archives in Washington, D.C. He lives in Bremo Bluff, Virginia.

Printed in the United States
by Baker & Taylor Publisher Services